THE ROUGH GUIDE to
Green Living

Duncan Clark

with contributions from
**Robert Henson &
Kevin Lindegaard**

D0277776

Credits

The Rough Guide to Green Living
Text & layout: Duncan Clark
Copyediting: Elizabeth Prochaska
Proofreading: Jason Freeman
Production: Rebecca Short

Rough Guides Reference
Series founder: Mark Ellingham
Editors: Kate Berens, Peter Buckley,
Matthew Milton, Ruth Tidball,
Tracy Hopkins, Joe Staines
Director: Andrew Lockett

Cover images: carbon footprint © Andrew Paterson / Alamy; tomatoes ©
Duncan Clark; electricity meter © Peter Titmuss / Alamy; train © Duncan Clark

Publishing information

This first edition published November 2009 by Rough Guides Ltd,
80 Strand, London WC2R 0RL
375 Hudson Street, New York 10014
Email: mail@roughguides.co.uk

Distributed by the Penguin Group
Penguin Books Ltd, 80 Strand, London WC2R 0RL
Penguin Putnam, Inc., 375 Hudson Street, NY 10014, USA
Penguin Group (Australia), 250 Camberwell Road, Camberwell, Victoria 3124, Australia
Penguin Books Canada Ltd, 90 Eglinton Avenue East, Toronto, Ontario, Canada M4P 2YE
Penguin Group (New Zealand), Cnr. Airborne and Rosedale Roads,
Albany, Auckland 1310, New Zealand

Typeset in Din, Minion and Myriad to an original design by Duncan Clark

336 pages; includes index

A catalogue record for this book is available from the British Library

Printed in Singapore by SNP Security Printing Pte Ltd

ISBN: 978-1-84836-107-2

1 3 5 7 9 8 6 4 2

Mixed Sources
Product group from well-managed
forests and other controlled sources
www.fsc.org   Cert no. SA-COC-001592
© 1996 Forest Stewardship Council
FSC

Contents

Part IV: Food

Part V: Shopping & services

Introduction

As society has started to come to terms with the massive implications of man-made climate change, environmental awareness
has reached an all-time high. It's clear now that unless we find
ways to slash global greenhouse gas emissions, then by the end
of the century much of the world might be radically less hospi-
table to humans and wildlife than it is today. As a result of this
increased awareness, everyone these days has an environmental
initiative or product to promote – politicians, supermarkets, even
oil companies. Magazines such as *Vanity Fair* produce special
green issues, supermarkets sell foods with labels showing their
carbon footprints, and motor shows are full of electric, hybrid
and hydrogen cars.

Despite this greening of seemingly every element of society, two things
have remained stubbornly constant. First, the world is still firmly on
a collision course with ecological disaster. As chapter one of this book
explains, when you take a cold look at the facts, it's fairly clear that only
a truly radical transformation of the way we use energy will be enough
to give us a decent chance of avoiding potentially devastating long-term
climate change.

The second thing that hasn't changed is that the world is still a hugely
confusing place for anyone who wants to live a more environmentally
friendly life. Contradictory information abounds and it's hard to identify
sensible advice among all the ill-informed hype and greenwash. Even
when you have the facts, there are plenty of tricky conundrums. What's
a greener choice, a hybrid car or an efficient diesel? Solar panels or extra
insulation? Organic food, local food or, even, industrially produced
food?

The main purpose of *The Rough Guide to Green Living* is to answer
these kinds of questions. It seeks to separate the facts from the fiction, and

to help readers get their environmental priorities in order. This last point is especially important given that many journalists seem intent on telling us that plastic bags, recycling and food miles are the most important environmental issues – which isn't even nearly true.

Before launching into the nitty-gritty of topics such as home energy, travel and food, the book takes a look at the bigger picture. Part one provides some background on climate change, carbon footprints and offsetting. It also touches on a more fundamental question: whether well-meaning individuals can actually make a difference anyway. If half the world consumes less energy and resources, won't that just make it cheaper for the other half to consume more? Such thoughts may not be very encouraging, but they help us understand that when it comes to a global environmental crisis, the real solutions have to be large-scale ones. This doesn't mean trying to live a greener life is a waste of time – not at all – but it does suggest we're likely to make a much bigger difference if we also engage on a higher level, via politics, pressure groups and other channels.

What's green?

Writing a guide to green living requires the setting of some boundaries about what does and doesn't constitute a "green" issue. This book takes a fairly broad view, so along with specifically environmental topics such as reducing your carbon emissions, you'll also find a certain amount of advice on topics such as fair trade and animal welfare. However, the book does focus first and foremost on straightforward environmental issues, and especially on all things related to climate change.

Climate change, or global warming, is the most urgent and wide-ranging environmental threat that humans have ever faced. Of course, there are many other pressing environmental issues, from species loss to dwindling fresh water supplies in many regions. Prioritizing climate change is not to play down these issues. On the contrary, global warming threatens to compound these other environmental threats as well as bringing about a wide range of impacts of its own.

Take species loss. There are many causes of this problem, including tropical deforestation for farmland and urban expansion, but the biggest cause of all, going forward, is likely to be the warming of the planet. Climate change could wipe out almost a third of all plant and animal species within decades, according to one landmark scientific study. The same is true of access to water: global agriculture may be rapidly running down ground-water levels, but climate change is having an even bigger impact, by expanding the world's dry areas and melting the glaciers which feed many important water sources.

In short, climate change is the pre-eminent environmental issue of our day, a planet-wide problem on a different scale to every other. That's why the majority of this book is about reducing greenhouse gas emissions, albeit with nods to other issues along the way.

Acknowledgements

This book has benefited from interviews, emails, images and advice from scores of people. My thanks go to all of them, as well as everyone at Rough Guides and Penguin, in particular Jonathan Buckley and Mark Ellingham for signing up my first Rough Guide green title back in 2003; Andrew Lockett for ushering through this volume; Joe Staines for years of press cuttings; Peter Buckley for picking up the slack on other projects; Ruth Tidball for her typically great editorial input. Special thanks to Robert Henson for allowing me to extract some text from his excellent *Rough Guide to Climate Change* (which I was lucky enough to edit) and to George Miller and Katharine Reeve for letting me reproduce the grow-your-own advice from their *Rough Guide to Food*. Finally, thanks to Profile and *Guardian* colleagues for giving me the opportunity to sharpen and deepen my environmental knowledge, and, above all, to Elizabeth Prochaska for advice, support, copyediting and lapsang souchong.

Part I

The big picture

Climate change
Your carbon footprint
Do individual actions matter?

Climate change

Key questions and answers

Given the importance of climate change to every chapter in this book, it's worth briefly going over some of the key facts and answering the most frequently asked questions about the nature of global warming. This chapter draws on text from *The Rough Guide to Climate Change*, which covers the science and symptoms of greenhouse warming – and the large-scale solutions – in much greater detail.

The basics

Is the planet really warming up?

In a word, yes. Independent teams of scientists have laboriously combed through more than a century's worth of temperature records (in the case of England, closer to 300 years' worth). These analyses all point to a rise of more than 0.7°C in the average surface air temperature of Earth over the last century. The first six years of the twenty-first century, along with 1998, were the hottest on record – and quite possibly warmer than any others in the past millennium.

Apart from what temperatures tell us, there's also a wealth of circumstantial evidence to bolster the case that Earth as a whole is warming up. Arctic sea ice has lost nearly half its average summer thickness since 1950, and by mid-century the ice may disappear completely each summer, perhaps for the first time in more than a million years. The growing season has lengthened across much of the Northern Hemisphere – hardly a crisis in itself, but a sign that temperatures are on the increase. At the same time, many mosquitoes, birds and other creatures are being observed at higher altitudes and latitudes, their change in habitats being driven by the increasing warmth.

But don't many experts claim that the science is uncertain?

There is plenty of uncertainty about details in the global-warming picture: exactly how much it will warm, the locations where rainfall will increase or decrease, and so forth. Some of this uncertainty is due to the complexity of the processes involved, and some of it is simply because we don't know how greenhouse emissions will change over time. But there's near-unanimous agreement that the global climate is already changing and that fossil fuels are at least partly to blame.

How could humans change the whole world's climate?

By adding enormous quantities of carbon dioxide and other greenhouse gases to the atmosphere over the last 150 years. These gases absorb heat that's radiated by Earth, but they release only part of that heat to space, which results in a warmer atmosphere.

The amount of greenhouse gas we add is staggering – in carbon dioxide alone, the total is more than thirty billion metric tonnes per year, which is more than four metric tonnes per person per year. And that gas goes into an atmosphere that's remarkably shallow. If you picture Earth as a soccer ball, the bulk of the atmosphere would be no thicker than a sheet of paper wrapped around that ball.

Even with these facts in mind, there's something inherently astounding about the idea that a few gases in the air could wreak havoc around the world. However, consider this: the eruption of a single major volcano – such as Krakatoa in 1883 – can throw enough material into the atmosphere to cool the global climate by more than 1°C for over a year. From that perspective, it's not so hard to understand how the millions of engines and furnaces spewing out greenhouse gases each day across the planet, year after year, could have a significant effect on the climate. (If automobiles spat out chunks of charcoal every few blocks in proportion to the invisible carbon dioxide they emit, the impact would be more obvious.)

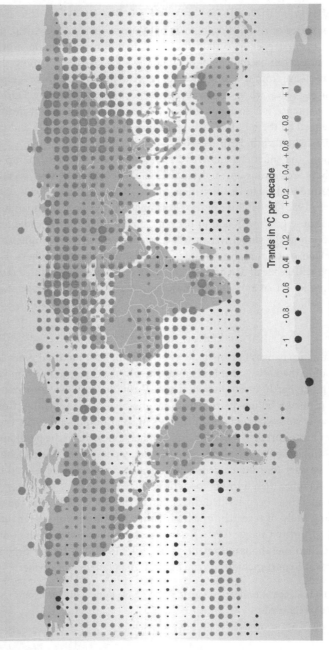

View of a warming world: a depiction of temperature change by region, 1976–2000

IPCC

Is a small temperature rise such a big deal?

While a degree or so of warming may not sound like such a big deal, the rise has been steeper in certain locations, including the Arctic, where small changes can become amplified into bigger ones. The warming also serves as a base from which heat waves become that much worse – especially in big cities, where the heat-island effect comes into play. During the most intense hot spells of summer, cities can be downright deadly, as evidenced by the hundreds who perished in Chicago in 1995 and the thousands who died in Paris in 2003.

Anyhow, the small changes we've seen so far could be just the tip of the iceberg. According to the best available science, the global average temperature is likely to rise anywhere from 1.1°C to 6.4°C by 2080–2099, relative to 1980–1999. This range reflects uncertainty about the quantities of greenhouse gases we'll add to the atmosphere in coming decades and also about how the global system will respond to those gases. Some parts of the planet, such as higher latitudes, will heat up more than others. The warming will also lead to a host of other concerns – from intensified rains to melting ice – that are liable to cause more havoc than the temperature rise in itself.

Couldn't the changes have natural causes?

The dramatic changes in climate we've seen in the past hundred years are not proof in themselves that humans are involved. As sceptics are fond of pointing out, Earth's atmosphere has gone through countless temperature swings in its 4.5 billion years. These are the results of everything from cataclysmic volcanic eruptions to changes in solar output and cyclic variations in Earth's orbit. The existence of climate upheavals in the past raises the question asked by naysayers as well as many people on the street: how can we be sure that the current warming isn't "natural" – ie caused by something other than burning fossil fuels?

That query has been tackled directly over the last decade or so by an increasing body of research, much of it through the Intergovernmental Panel on Climate Change (IPCC), a unique team that draws on the work of more than one thousand scientists. Back in 1995, the IPCC's Second Assessment Report included a sentence that made news worldwide:

> "The balance of evidence suggests a discernible human influence on global climate."

By 2001, when the IPCC issued its third major report, the picture had sharpened further:

> "There is new and stronger evidence that most of the warming observed over the last 50 years is attributable to human activities."

And in its fourth report (2007), the IPCC spoke even more strongly:

> "Human-induced warming of the climate system is widespread."

One way in which scientists attribute climate change to greenhouse gases is by looking at the signature of that change and comparing it to what you'd expect from non-greenhouse causes. For example, over the past several decades, Earth's surface air temperature has warmed most strongly near the poles and at night. That pattern is consistent with the projections of computer models that incorporate rises in greenhouse gases. However, the pattern agrees less well with the warming that might be produced by other causes, including natural variations in Earth's temperature and solar activity.

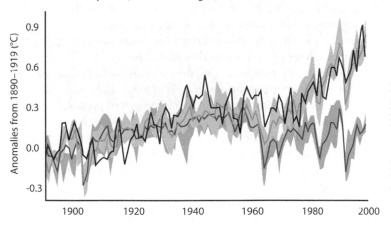

The temperature changes observed in the last hundred years fit nearly exactly with sophisticated computer simulations – though only when man-made greenhouse warming is included in the models.

As computer models have grown more complex, they've been able to incorporate more components of climate. This allows scientists to tease out the ways in which individual processes helped shape the course of the last century's warm-up. One such study, conducted at the US National Center for Atmospheric Research, examined five different factors: volcanoes, sulphate aerosol pollution, solar activity, greenhouse gases and ozone depletion. Each factor had a distinct influence. The eruption of Mount Pinatubo in 1991 helped cool global climate for several years. Sulphate pollution peaked in the middle of the twentieth century, between World War II and the advent of environmentalism, and it may have helped produce the slight cool-down observed in the middle of the twentieth century. Small ups and downs in solar output probably shaped the early-century warming and perhaps the mid-century cooling as well. However, the sun can't account for the pronounced warming evident since the 1970s. The bottom line is that the model couldn't reproduce the most recent warming trend unless it included greenhouse gases.

Couldn't some undiscovered phenomenon be to blame?

Although many people would love to find a "natural" phenomenon to blame for our warming planet, such as the relationship between clouds and cosmic rays, it's growing extremely unlikely that a suitable candidate will emerge. Even if it did, it would beg a difficult question: if some newly discovered factor can account for the climate change we've observed, then why aren't carbon dioxide and the other greenhouse gases producing the warming that basic physics tells us they should be?

And there's another catch. Any mystery process could just as easily be a cooling as a warming agent, and if it were to suddenly wane, it could leave us with even greater warming than we imagined possible. Trusting in the unknown, then, is a double-edged sword. As such, most scientists in the trenches trust in Occam's razor, the durable rule credited to the medieval English logician and friar William of Occam: "One should not increase, beyond what is necessary, the number of entities required to explain anything."

How do the rainforests fit into the picture?

The destruction of rainforests across the tropics is a significant contributor to climate change, accounting for roughly a fifth of recent human-produced CO_2 emissions. Tropical forests hold nearly half of the carbon present in vegetation around the world. When they're burned to clear land, the trees, soils and undergrowth release CO_2. Even if the land is eventually abandoned, allowing the forest to regrow, it would take decades for nature to reconcile the balance sheet through the growth of replacement trees that pull carbon dioxide out of the air. In addition to the CO_2 from the fires, bacteria in the newly exposed soil may release more than twice the usual amount of another greenhouse gas, nitrous oxide, for at least two years. Brazil's National Institute for Amazon Research estimates that deforestation puts four times more carbon into the atmosphere than the nation's fossil-fuel burning does.

The impacts

Is global warming necessarily a bad thing?

Whether climate change is bad, good or neutral depends on your perspec tive. Some regions – and some species – may benefit, at least in the short term, but many more will suffer intense problems and upheavals. And some of the potential impacts, such as a major sea-level rise, increased flooding and droughts, more major hurricanes and many species being consigned to extinction, are bad news from almost any perspective.

Perhaps the more pertinent question is whether the people and institu tions responsible for producing greenhouse gases will bear the impacts of their choices, or whether others will – including those who had no say in the matter. Indeed, people in the poorest parts of the world – such as Africa – will generally be least equipped to deal with climate change, even if the changes are no worse there than elsewhere. Yet those regions have released only a small fraction of the gases that are causing the changes.

How big will the human cost be?

Quantifying the human cost of climate change is difficult. Weather-related disasters kill thousands of people each year, regardless of long-term changes in the climate. Furthermore, many of the projected impacts of global warming on society are the combined effects of climate change and population growth. For this reason, it's hard to separate out how much of the current and future impacts are due to each factor, though some experts have tried. A recent report by the Global Humanitarian Forum, a think tank headed by former UN secretary general Kofi Annan, estimated that climate change is already responsible for around 300,000 deaths per year, with a further 300 million people seriously affected.

In the decades to come, the warming of the planet and the resulting rise in sea level will likely begin to force people away from some coastlines. Low-lying islands are already vulnerable, and entire cities could eventually be at risk. The implications are especially sobering for countries such as Bangladesh, where millions of people live on land that may be inundated before the century is out. Another concern is moisture – both too much and too little. In many areas rain appears to be falling in shorter but heavier deluges conducive to flooding. However, drought also seems to be becoming more prevalent. Changes in the timing of rainfall and runoff could seriously complicate efforts to ensure clean water for growing populations, especially in the developing world. Warming temperatures may also facilitate the spread of vector-borne diseases such as malaria and dengue fever. A report published in 2009 in the *Lancet* medical journal named climate change as the biggest threat to human health in the twenty-first century.

Among those scientists prepared to make longer-term predictions, some – such as James Lovelock, the inventor of Gaia theory – foresee unchecked climate change causing a total meltdown of human civilization. "Billions of us will die and the few breeding pairs of people that survive will be in the Arctic where the climate remains tolerable", Lovelock has written. It should be noted, though, that many other scientists believe this kind of extreme pessimism is both unfounded and unhelpful.

How will food production be affected?

That depends on where the farming or ranching is done, and on how effectively mankind manages to reduce emissions. One study commissioned by the UN suggests that global agricultural productivity is likely to go up over the next century, thanks to the fertilizing effect of the extra CO_2 in the atmosphere and to now-barren regions becoming warm enough to bear crops. Unfortunately, any benefit will be limited to the developed world: productivity in the tropics, home to millions of the most vulnerable subsistence farmers (and to much of the expected rise in population in the coming decades) is predicted to fall. A decline in global food production can be expected, too, if the world becomes three degrees hotter than it was in the pre-industrial age, according to the most recent IPCC report.

Will New York and London really end up under water?

Not right away, but it may be only a matter of time. In its 2007 report, the IPCC projected that sea level will rise anywhere from 180 to 590mm by

2090–2100. However, this estimate excluded some key uncertainties about how quickly warming will melt land-based ice, and more recent research has suggested that this century's rise could be more than a metre. A rise on that scale would cause major problems in many regions – especially when hurricanes and coastal storms are taken into account.

In the longer term, if emissions continue to rise unabated through this century, the Greenland and/or West Antarctica ice sheets could be thrown into an unstoppable melting cycle that would raise sea level by more than 7m (23ft) each. This process would take some time to unfold – probably a few centuries, although nobody can pin it down at this point – but should it come to pass, many of the world's most beloved and historic cities would be hard-pressed to survive.

How will the economy be affected?

There is a divergence of views regarding the impact of climate change – and indeed emissions cuts – on national and international economies in the coming century. The most widely cited research into this question was carried out on behalf of the UK government by the economist Nicholas Stern. *The Stern Review* concluded that global warming could reduce global GDP by 20% within the current century and recommended spending 1% of global GDP each year – starting now – on reducing emissions to help counter that threat. Stern has subsequently admitted that his report underestimated the speed, scale and likelihood of some serious climate impacts and he now advocates spending 2% of GDP on mitigation.

Other economists have questioned Stern's methodology, however, and have estimated either higher or lower figures for the costs of avoiding and ignoring climate change.

And wildlife?

Because the climate is expected to change quite rapidly from an evolutionary point of view, we can expect major shocks to some ecosystems – especially in the Arctic – and possibly a wholesale loss of species. According to a 2004 study led by Chris Thomas of the University of Leeds and published in the journal *Nature*, climate change between now and 2050 may commit as many as 37% of all species to eventual extinction – a greater impact than that from global habitat loss due to human land use. Similar figures emerged from the 2007 IPCC report, which pegs the percentage of plant and animal species that are at risk from a temperature rise of 1.5–2.5°C (a level that now seems almost inevitable) at 20–30%.

The causes

What are the sources of all the greenhouse gases?

At present, almost every human activity results either directly or indirectly in the release of greenhouse gases into the atmosphere. That's because most activities currently rely directly or indirectly on energy provided by fossil fuels – coal, oil and natural gas. These fuels, which create CO_2 when they're burned, power everything from the tractors on our farms through to the light bulbs in our homes. Aside from fossil fuels, CO_2 is also released by deforestation, and other greenhouse gases – such as methane and nitrous oxide – are emitted by agriculture, rotting waste and various industrial processes. The box overleaf describes the main greenhouse gases and their sources, and the pie chart below shows one breakdown of global emissions into various sectors of the world economy.

As some climate change sceptics like to point out, even our breathing puts CO_2 into the air. In this case, however, there's no need to feel bad. The carbon we exhale was previously plucked from the atmosphere by plants – either those which we ate directly or those consumed by animals being reared for meat. Biologically, then, we're basically carbon neutral. It's our energy use, industries and agriculture that are the problem.

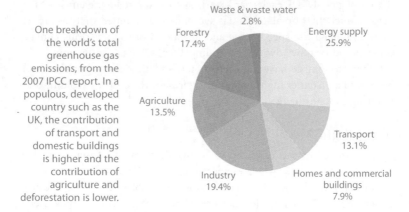

One breakdown of the world's total greenhouse gas emissions, from the 2007 IPCC report. In a populous, developed country such as the UK, the contribution of transport and domestic buildings is higher and the contribution of agriculture and deforestation is lower.

Waste & waste water 2.8%
Forestry 17.4%
Energy supply 25.9%
Agriculture 13.5%
Transport 13.1%
Industry 19.4%
Homes and commercial buildings 7.9%

Which countries are emitting the most greenhouse gases?

There are various ways to compare the greenhouse output of nations. The most obvious metric is current national CO_2 emissions per year. By that measure, China is in first place, having recently overtaken the US. Each of those two countries accounts for around a fifth of global emissions. In third, fourth and fifth place are India, Russia and Germany. The UK comes in at number eight, with around 2% of total emissions.

Looking at current total emissions can be misleading, however. For one thing, since CO_2 remains in the atmosphere for around a century, it's important to also consider historical emissions. This gives a more accurate impression of the total responsibility of each country for the predicament the world finds itself in today. By this measure, the US is the clearer leader, with around 30% of the fossil-fuel CO_2 ever released. Russia and China come next, with around 8% each, followed by Germany, with 7%, and the UK, with 6%.

Even this doesn't give the full picture, however, since countries vary widely in the number of people they contain. To understand the responsibility of individuals within nations we need to consider the emissions per person. Looked at this way, developing countries shoot down the ranking, as their emissions are split between more than a third of the world population. The US remains very high, though is overtaken by countries such as Australia, where the per capita emissions are around five times higher than those in China, and around 25 times those in India. We'll look at the typical emissions of UK and US citizens in the following chapter.

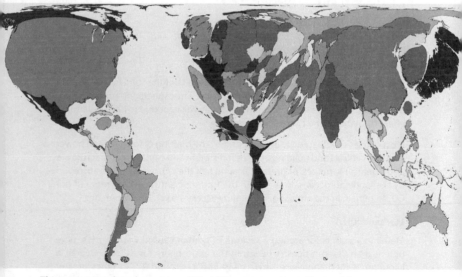

This map, reproduced courtesy of WorldMapper.org, shows the world's countries resized according to their greenhouse emissions.

Greenhouse gases and their sources

The temperature of any planet is determined partly by the gases in its atmosphere. So-called "greenhouse gases" act like a blanket, absorbing heat that would otherwise escape to space, thereby warming up the planet's surface. This phenomenon is known as the greenhouse effect.

The Earth has always had greenhouse gases in its atmosphere. If it had none, the difference between daytime and night-time temperatures would be extreme. Overall, the planet would be much colder, with an average temperature of -18ºC, as opposed to the much more habitable 14ºC we have today.

Over the past two centuries, human activities have been increasing the concentration of various greenhouse gases in the atmosphere. Following is a brief summary of the most important gases and their sources.

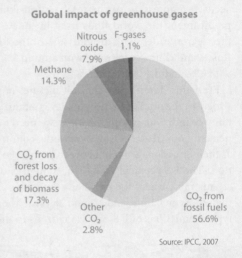

Global impact of greenhouse gases

Nitrous oxide 7.9%

F-gases 1.1%

Methane 14.3%

CO_2 from forest loss and decay of biomass 17.3%

Other CO_2 2.8%

CO_2 from fossil fuels 56.6%

Source: IPCC, 2007

Carbon dioxide (CO_2)

Carbon dioxide accounts for around three quarters of the total human impact on the climate. The gas typically persists in the atmosphere for around a hundred years, which means emissions from today will continue to have a warming effect throughout the twenty-first century.

Most of the CO_2 released today is emitted by the burning of fossil fuels in power stations, vehicles and buildings, but deforestation and cement manufacture are also significant sources. At present, around half the CO_2 we release is absorbed by the land and oceans. However, in the future, scientists expect more of our CO_2 to remain in the atmosphere, thereby accelerating global warming.

Methane (CH_4)

Methane accounts for around a seventh of human-caused warming. It's more potent than CO_2 but doesn't persist in the atmosphere for as long. Overall, a kilogram of methane causes a warming impact around 20–25 times greater than a kilogram of CO_2.

Significant sources of methane include sheep and cows (which belch the gas from their ruminant stomach systems), rice paddies, landfill sites and the production of fossil fuels. Methane is basically the same as the natural gas we use for cooking and heating, though once it's burned on a stove or a boiler, it combines with oxygen to become CO_2.

Nitrous oxide (N_2O)

Nitrous oxide accounts for around 8% of human emissions. Relatively small amounts of the gas are emitted, but each kilogram causes a warming effect hundreds of times greater than a kilo of CO_2. N_2O is released naturally by bacteria living in soils, but humans substantially increase the rate at which this happens through the addition of fertilizers. Livestock manure is a particularly potent source of this greenhouse gas, a fact that provides one environmental argument for reducing our intake of animal products (see p.212). Other sources of N_2O include power plants and certain industrial processes.

"F-gases"

The F-gases are a family of synthetic chemicals used as solvents, refrigerants and aerosol propellants, among other things. They include hydrofluorocarbons, perfluorinated carbons and sulphur hexafluoride. F-gases are among the most powerful of all greenhouse gases – thousands of times more potent than CO_2. Thankfully, they're not emitted in particularly large quantities.

Other greenhouse gases: water vapour (H_2O) and ozone (O_3)

In the case of the above four greenhouse gases, the formula is pretty simple: the more we release, the more warming we create. But there are two other significant greenhouse gases whose relationship both to human activity and global warming is somewhat more complex. The first is water. Taking the form of clouds and humidity, water vapour makes up roughly 1% of the Earth's atmosphere. It accounts for the majority of the "natural" greenhouse effect and has a greater warming impact than all the human-produced gases put together.

Water vapour is evaporated from the seas and land as part of the Earth's water cycle. Human-produced global warming may accelerate this process, since more heat would lead to more evaporation. This, in turn, would cause more warming – and so on. Humans can also make some small changes to the cloud system directly. Plane contrails can lead to cirrus clouds, which many experts believe increases their climate impact (see p.170).

The role of ozone is similarly complicated. This unstable version of oxygen exists naturally in the upper atmosphere (the ozone layer) but can also be created by the reaction between sunlight and ground-level pollution. Ozone itself is a greenhouse gas, but by reacting with other gases it can also shorten the livespan of other greenhouse gases.

Solutions

Can we still fix climate change?

Probably, yes, though it won't be easy and the world will need to act very quickly. The climate system includes a number of positive-feedback processes that are expected to significantly amplify the warming caused directly by humans. For example, as ice melts, it exposes sea and rock that will tend to absorb more heat than the reflective ice did. Similarly, as soils warm, they emit more nitrous oxide. The concern is that if we push the climate system too far, we'll reach a "tipping point" at which these positive feedbacks take over and make global warming both more intense and irreversible.

Since each positive feedback has its own triggering mechanism, there is no single temperature agreed upon as a tipping point for Earth as a whole. However, scientists, governments and activists have worked to identify useful targets. One goal adopted by the European Union, as well as many environmental groups, is to limit global temperature rise to 2°C over pre-industrial levels. But that ceiling looks increasingly unrealistic – we're already close to 40% of the way there, and only the lower fringes of the latest IPCC projections keep us below the 2°C threshold by the century's end. According to a recent report from the UK government's independent Committee on Climate Change, we'd need to see global emissions peak and begin to fall by 2014 in order to have even a fifty percent chance of avoiding the two-degrees threshold.

Another approach to forming a global target is to set a stabilization level of greenhouse gases – a maximum concentration to be allowed in the atmosphere. Currently, we're on 380 parts per million, as compared to 270–280ppm in pre-industrial times. Worryingly, many scientists now believe that 380 is *already* above the safe limit, and that we need to somehow reduce the carbon dioxide and other greenhouse gases in the atmosphere down to around 350ppm. To do that quickly would probably mean slashing greenhouse emissions almost to zero while at the same time finding ways to increase the amount of CO_2 soaked up by the seas and soils, or to capture and store the gases underground. Unfortunately, it seems extremely unlikely that policymakers around the world will show the vision, leadership and diplomacy to execute such a radical plan.

In other words, we could still avoid irreversible climate change, but it would require either an unprecedented global effort to change our energy sources and lifestyles or a lot of luck in the form of a climate system that turns out to be less sensitive than most experts believe. Or both.

Kyoto and beyond

At the 1992 Earth Summit almost all nations signed the UN Framework Convention on Climate Change, which pledges to stabilize greenhouse-gas concentrations at a level sufficient to "prevent dangerous anthropogenic interference with the climate". Five years later, in Kyoto, Japan, the various signatories finalized a protocol to give teeth to that pledge. The Kyoto Protocol mandated targets for emissions reductions for developed nations averaging 5.2% by 2012, relative to 1990 levels.

Over the following years, many countries ratified the treaty, though some – notably the US and Australia – refused, on the grounds that they shouldn't have to make cuts unless developing countries promised to do the same. Kyoto finally became law in 2005, when the proportion of the developed world's CO_2 emissions represented by countries that had ratified the treaty reached the required 55%. By that time, emissions were climbing in many countries and rocketing in China and various other developing nations.

Kyoto is, at best, just a first step towards a meaningful solution to climate change. Even if all nations met their targets for 2012, then the result would be just a small dip in total emissions. The reality is even worse, as few major emitters are likely to hit their targets. The big question, then, is what will follow on from Kyoto in 2012. Policymakers will attempt to answer this question at a summit in Copenhagen in December 2009. Whether they succeed in hammering out a new global deal ambitious enough to arrest global warming this century remains to be seen.

Might recession or peak oil solve the problem?

Carbon emissions tend to move in lockstep with economic activity so it's true that an economic downturn of the sort we've seen since 2008 will have some impact on the world's carbon output. However, even a very severe global recession would be unlikely to reduce global emissions by more than a few percent, and that benefit would almost certainly be more than offset by falling investment in alternative energy. All told, then, there's little hope for an environmental silver lining in the slump.

The same is true for "peak oil" – the point when oil production reaches its final maximum and begins a terminal decline. If the peak arrives in the next few years, as some experts expect, then it might just cause an economic catastrophe severe enough to dint global emissions for some time. That, however, is an unknown. The big question is which fuel sources the world turns to as a substitute for the dwindling oil supplies. If renewables such as solar and wind manage to scale up in time to fill the energy gap, then there will be plenty of reason for optimism. But if the world falls back on coal, then we'll be staring disaster in the face.

What can I do, and will it make any difference?

As the rest of this book shows, there's plenty that any individual can do to help reduce emissions. Chapter two will help you understand your carbon footprint, while chapter three explains why it's also important to see the bigger picture and to consider getting involved in other ways. For more detailed practical advice, flick ahead to the sections on home, food and travel that make up the bulk of the book.

Your carbon footprint

Understanding and measuring your impact on the climate

A carbon footprint is a measure of the total greenhouse gas emissions caused by an activity, item, person, business or country. If "footprint" seems an odd choice of metaphor for the release of gases into the air, that's because the concept of the carbon footprint can be traced back to an earlier, broader, environmental measure: the ecological footprint. This phrase, coined in the early 1990s by Canadian academics William Rees and Mathis Wackernagel, describes the area of the Earth's surface required to supply the food and goods for an individual's lifestyle. We'll come back to ecological footprints later in this chapter, but for now, let's focus on carbon.

Understanding carbon footprints

The concept of a carbon footprint – the total global warming impact of something – is simple enough. For the uninitiated, however, the numbers can be close to meaningless. Most of us aren't particularly good at thinking in terms of units of mass – such as kilograms and tonnes – at the best of times. When the thing being measured happens to be a colourless, odourless gas, things can get hopelessly confusing and abstract.

To help you make sense of different weights of carbon dioxide, following are some rough pointers. If you can't get used to grams and kilograms, however, don't worry too much: in this book, carbon footprints are usually expressed not just in terms of weight but also in terms of more easily graspable equivalents such as the miles driven in an average car.

▶ **10g of CO_2** weighs as much as a pound coin and would fill a shoe box. Equivalent to driving fifty metres.

▶ **100g of CO_2** weighs as much as a mobile phone and would fill a household oven. Equivalent to driving five hundred metres.

▶ **1kg of CO_2** weighs as much as a bag of sugar and would fill a spherical balloon one metre across. Equivalent to driving five kilometres.

▶ **10kg of CO_2** weighs as much as a sack of onions and would fill a family car. Equivalent to driving fifty kilometres.

▶ **100kg of CO_2** weighs as much as a large man and would fill a single-decker bus. Equivalent to driving five hundred kilometres.

▶ **1 tonne of CO_2** weighs as much as a car and would fill a large detached house. Equivalent to driving around five thousand kilometres.

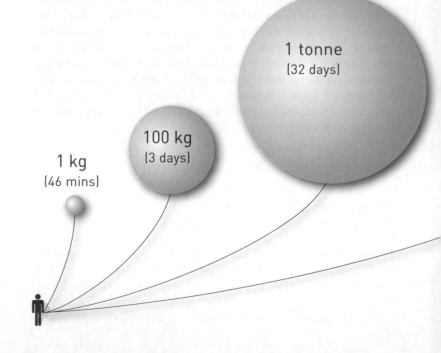

11.6 tonnes
(1 year)

Spherical balloons showing the volume of various weights of CO_2, with a person of average height depicted for scale. The time specified on each balloon shows roughly how long it takes for each UK citizen to generate that amount of CO_2 through all their purchases, travel, energy use and activities. (If other greenhouse gases such as methane and nitrous oxide were included the balloons would be even bigger for each time period.)

Carbon equivalent

Despite the name, "carbon footprint" refers to *all* the greenhouse gases caused by an item or activity – not just the CO_2. For example, the carbon footprint of beef includes the methane belched up by the cattle. Things would get confusing if carbon footprints were expressed as a certain amount of one gas and a certain amount of another. To avoid this, and to keep things simple, footprints are expressed as "carbon equivalent" – that is, the amount of CO_2 alone that it would take to create the same overall warming impact. For this reason, you'll often see the unit "CO_2e" used as the measure of a carbon footprint, though very often (in this book and elsewhere) the e is left off.

Why carbon footprints aren't an exact science

The aim of measuring the carbon footprint of something – whether that something is a simple bag of sugar or a person's entire life – is to understand all the climate-changing emissions associated with it. That sounds simple enough, and for some activities, such as burning a litre of petrol in a car, we can say how much CO_2 was caused to a fair degree of accuracy.

For most products and activities, however, accurately measuring the carbon footprint is very difficult. Even if you were trying to work out the footprint of something as simple as a bag of sugar, there would be a huge range of energy flows to factor in – from the oil used by the farmer's machinery and the natural gas used to produce the fertilizer through to the fuels used to process, pack, transport and distribute the crop. Greenhouse gases might also be released by the soils during the farming and by the crop wastes as they break down.

In reality, to get accurate information about all these factors would usually be impossible, but even if you *did* have all the data to hand, there would be still many difficult conundrums left to resolve. To pick one of countless possible examples, should the footprint of the sugar include the emissions from the petrol consumed by the farm labourers driving to and from work? Or how about the CO_2 released into the atmosphere five years ago when a patch of virgin rainforest was cleared to make room for the sugar plantation? There are no simple answers to questions such as these, so carbon footprints are always based on certain assumptions and simplifications.

Your carbon footprint

If it's so complicated to work out the carbon footprint of a bag of sugar, then wouldn't it be impossible to accurately work out the footprint of an entire lifestyle, with all the thousand of products and activities it includes? In short, yes: it's practically impossible to accurately work out your true carbon footprint from first principles. Nonetheless, it *is* perfectly possible to get a rough sense of your footprint – and to see where you stand in relation to others in your country and the wider world, and in relation to what experts describe as a sustainable limit.

A good start is to play with a carbon calculator. These simple online tools allow you to calculate how much carbon various activities in your life generate. There are scores of calculators out there, and they vary widely in accuracy and scope. Many calculators, such as those offered by the various carbon-offset companies (see p.314), simply let you tot up the emissions caused by a few high-emission activities, such as driving, flying and heating your home.

These kinds of sites, though useful, miss a lot out. A more comprehensive analysis should include other emissions sources such as the production and transport of all the food and consumer goods we buy. Bags of sugar, for example. And to give the full picture, we'd need to include all the government and private buildings and services we rely upon – from hospitals and schools through to the police and garbage collection.

There are one or two carbon calculators that attempt to cover some of these kinds of emissions, such as WattzOn (see p.28). However, it's often more informative to simply look at the average data for your entire country. The charts overleaf give one breakdown of the carbon footprint of the average UK and US residents, with an average figure for Africa provided for comparison. These figures take the entire emissions of the UK and US and simply divide them by the number of citizens. This way a share of all emissions are included – from house building to agriculture.

Outsourced emissions: whose carbon is it?

The typical footprint figures shown overleaf include not just the average person's share of emissions on p.25 within their own country. They also include emissions created overseas in the farms and factories producing the goods that we consume. To use the jargon they show footprints based on the "consumption" of CO_2, not just the "production".

Environmentalists tend to argue that a consumption footprint is the only meaningful measure of a person's contribution to global warming. It would be wrong, they argue, to expect China to accept responsibility for the emissions created by a Chinese factory producing toys for British children. The West has "outsourced" or "offshored" most of its manufacturing industries but that doesn't reduce our responsibility for the environmental impact of those industries. Studies by the Stockholm Environment Institute and economist Dieter Helm have shown that as much as 35% of the footprint of a UK citizen is created overseas.

On the other hand, some policymakers argue that manufacturing nations such as China benefit economically from producing goods for export, and therefore they should be held at least partly responsible for those emissions. In the words of Elliott Morley, chair of the UK Parliament's Energy and Climate Change Committee, "Everyone has some responsibility. It is true that UK emissions have been offshored. But the UK has paid a price in terms of lost jobs, while China has benefited from job creation."

Reducing your footprint

The rest of this book is largely about how to reduce your footprint. As the charts on the opposite page show, about half of the average UK or US citizen's emissions are accounted for by home energy and travel, so these areas are the subject of the next part of this book. Air travel is particularly important – if you're a frequent flier then your carbon footprint will probably be *much* higher than the average for your country, not to mention the global average.

But what about the other half of our emissions? These are caused by the production and transportation of the food and goods we buy, the construction of our homes, offices and roads, and the services we use. We can reduce some of these emissions by avoiding unnecessary purchases, favouring low-carbon food and other goods, and by recycling – all things discussed later in the book.

Another approach is to favour brands, shops and services that have sought to reduce their own emissions, and avoid those which have lobbied against mandated cuts in greenhouse-gas emissions. Until fairly recently, the latter category included most of the mainstream business community, especially in the US. (For example, in 2001 the CEO of the US Chamber of Commerce, which represents three million businesses, wrote a letter to President Bush claiming that "the Kyoto Protocol is a flawed treaty that

ONE BREAKDOWN OF THE TYPICAL PERSON'S CO_2 EMISSIONS

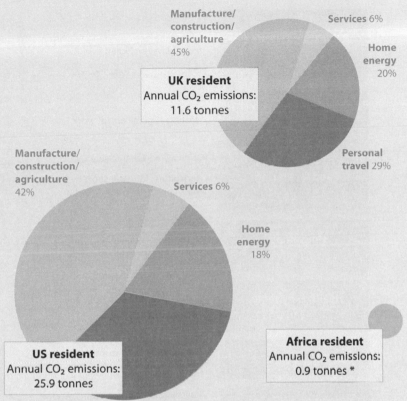

Manufacture/construction/agriculture 45%

Services 6%

Home energy 20%

UK resident
Annual CO_2 emissions:
11.6 tonnes

Personal travel 29%

Manufacture/construction/agriculture 42%

Services 6%

Home energy 18%

US resident
Annual CO_2 emissions:
25.9 tonnes

Africa resident
Annual CO_2 emissions:
0.9 tonnes *

Personal travel 34%

These diagrams, based on data from BestFootForward.com, show breakdowns of the typical CO_2 emissions per capita in the US and the UK, with a typical African citizen provided for comparison. The footprint figure for each resident is reached by simply dividing the total for each country or region by the population. The figures shown here are higher than the official government statistics because they include international transport emissions and imported goods, both of which are ignored under the Kyoto Protocol. Note, however, that these figures don't reflect total carbon footprints as they don't include methane or nitrous oxide. If they did, the figures would shoot up substantially, with the extra emissions allocated mainly to agriculture and manufacture. However, accurate figures for the methane and nitrous oxide emissions caused by the growing of our foods overseas are not easily available.

* The figure for Africa excludes aviation and imported goods, as the data isn't available, though these additional emissions would be minimal

ANOTHER BREAKDOWN OF THE TYPICAL PERSON'S CO₂ EMISSIONS

Recreation & leisure
1.93 tonnes

Everything from watching television to holidays (excluding flying)

Space heating
1.47 tons

Heating at home and at work

Food & catering
1.37 tonnes

Agriculture, food transport by road, cooking and restaurants

Household
1.36 tonnes

Lighting, home building, decoration, gardening, DIY, etc

Health & hygiene
1.33 tonnes

Bathing, showering, washing and health services

Clothing 9.8 tonnes

Production, road transportation, retail and washing/drying of clothes and shoes

Commuting 0.80 tonnes

Travelling to and from work by car or public transport

Aviation 0.67 tonnes

Leisure and business travel, plus air-freight

Education 0.48 tonnes

Schools and school travel, plus books and newspapers

Government & defence (0.29 tonnes)

Communication (1.6 tonnes)

This analysis of the typical UK citizen's carbon footprint was adapted from *The Carbon Emissions Generated in all that we Consume* by the Carbon Trust. Again, national CO₂ emissions are divided equally among the citizens but this break down shows a wider range of activities. So driving, for example, is split between leisure and commuting. As before, the figures in this diagram focus only on CO₂; if other greenhouse gases and impacts were included aviation and food emissions would grow substantially.

is not in the US interest".) Today, however, most major companies have recognized the problem and some have managed to cut their emissions considerably. Rewarding these businesses with our custom would seem sensible.

Some emissions sources are essentially beyond our control. We all benefit directly or indirectly from schools, hospitals and many other government services, but the only simple way to tackle these emissions – other than by deciding to have fewer children – is to engage as a citizen rather than as a consumer. This is discussed in chapter three.

What's a sustainable carbon footprint?

When seeking to reduce your emissions, it obviously would help to have a target footprint in mind. So how much CO_2 could each of us safely emit without risking the health of the planet? As is so common with climate change, there's no simple answer to this question. The sustainable foot print for each person on the planet depends on factors such as the world population (which is still growing fast) and how much risk of runaway global warming we're prepared to accept. Some scientists claim we're already well into the danger zone, which would imply that the only way to be truly sustainable would be to reduce your carbon footprint to zero. Other experts claim we still have a decade or so to solve the problem, and that if everyone reduced their footprint to around two tonnes – a cut of approximately 80–90% for a UK citizen, depending on how you measure it – then we'd be well on the way to a sustainable future.

Can you be rich and green?

In one sense, the wealthier members of society have the biggest capacity to solve environmental problems. If you can afford to overhaul your home with an eco-renovation and scrap your old car in favour of an electric model, then you'll have the potential to effect more change than someone with a limited budget focusing on the smaller actions. As a rule, however, there's a strong correlation between earnings and carbon footprint. Indeed, there's a certain irony to the fact that the relatively wealthy middle-class people who try hardest to be green are very often those who account for a disproportionate slice of the problem. The middle classes are good at shunning carrier bags, recycling and bemoaning the modern world's excessively consumer culture, but not necessarily so good at actually reducing their footprints.

Various studies in the UK, US and Australia have shown that relatively well-off, educated individuals cause as much as twice the carbon emissions as the average person in their country. This isn't surprising when you consider that the middle and upper classes take the largest number of foreign holidays, live in the biggest homes, worry least about their gas and electricity expenditure, and have the largest disposable income for everything from exotic fruit to laptops. Ultimately, almost every pound or dollar spent generates some carbon emissions, and while it's possible in theory to spend a lot of money on climate-friendly items such as solar panels, in reality few people manage to have both a high income and a low footprint.

Online carbon and energy tools

Many websites offer tools to help you measure, manage, track and reduce your carbon footprint. As previously mentioned, most focus solely on direct emissions – such as driving and flying – while ignoring indirect ones such as purchases and government services. Nonetheless, the few sites listed below are definitely worth exploring. For quickly calculating the emissions of a specific activity, such as a flight, you might be better using one of the carbon offset websites listed on p.314.

WattzOn www.wattzon.com

WattzOn focuses on energy – hence the "Watt" in the name – rather than carbon footprints as such. But it will tell you the total amount of energy required to support your lifestyle – and the CO_2 currently emitted in the production of that energy.

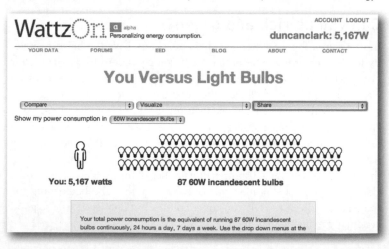

The interface is excellent, and the results are easy to understand thanks to a comparison system which lets you visualize your energy footprint in terms of tangible metrics such as a number of constantly illuminated lightbulbs. Check out the EED (Embedded Energy Database) to get a sense of the energy and emissions created by the manufacture and transport of a wide range of items.

The Carbon Account www.thecarbonaccount.com

This nicely designed website lets you track your direct emissions over time. You log in and update your "carbon account" with data such as electricity and gas meter readings, petrol consumption and flights taken. The results can be viewed on a graph, helping you see how your emissions are changing over time.

Bloom www.bbc.co.uk/bloom

This BBC climate change site offers a range of green-living background guides (some of them written by the author of this book) as well as detailed information on the carbon impact of seventy five different actions, from installing low-energy bulbs to adding a green roof to your home. If you find the Flash animations irritating, look for the link for the static version of the site.

Carbon Trust Calculator www.carbontrust.co.uk/publicsites/CFCalculator

The Carbon Trust was set up by the UK government to help businesses reduce emissions. Their online carbon calculator will be useful for anyone who runs a small business and wants to get a rough sense of its direct carbon footprint.

Ways to think about costs and savings

By reducing your carbon footprint, you'll usually be reducing your energy consumption, which should save you money. In some cases, such as turning down the heating or not flying long-haul on holiday, you won't need to spend anything upfront in order to make the savings. In many cases, however, you'll need to balance up savings recouped in the medium or long term with an initial outlay – anything from the price of buying an eco gadget to the cost of having your cavity walls insulated. There are various ways to think about this kind of calculation:

▶ **Payback period** This tells you the period of time after which you can expect the savings to have cancelled out the cost of investment. If you need to spend, say, £200 on improving your insulation, and you expect this would reduce your energy bills by £20 a year, then the payback period would be ten years.

▶ **Return on investment** The problem with thinking in terms of payback periods is that it doesn't consider what else you might have done with the money. For that reason, it's sometimes more instructive to think about the return on investment. To continue with the same example, by slashing your bills by £20 per year, the £200 investment in insulation would be earning a 10% return – a much better rate than you'd typically get if you left the money in a bank account. Of course, unlike with a bank account you can't withdraw the money at a future date (unless you sell the home, which hopefully will be worth a little more thanks to the insulation).

▶ **Cost per tonne of CO_2 saved** Payback periods and returns on investment focus purely on financial costs and benefits. To factor in the ecological benefits, it's useful to think in terms of the cost per unit of CO_2 saved. If the extra insulation reduced the heating bills in your home by half a tonne per year, then the cost per tonne of CO_2 would be very low. Even if the home was demolished in fifty years' time, then the overall cost would have been as little as £4 per tonne. Alternatively, you could spend the same £200 on a new, more efficient dishwasher to reduce your electricity consumption. In this case, you might find that the savings would only be around 100kg (a tenth of a tonne) of CO_2 per year. So if the dishwasher lasted ten years, the savings would have cost around £200 per tonne. In other words, you'd get much less ecological benefit for your money with the dishwasher than you would with the insulation.

Ecological footprints

The concept of a carbon footprint grew out of a longer-established and broader idea: the ecological footprint. This is a measure of our consumption levels in terms of the total area of the Earth's surface needed to support our individual existence. This area – measured in hectares of average productivity – includes the space for growing crops, grazing animals, harvesting timber, catching fish, accommodating infrastructure, absorbing carbon dioxide emissions and storing waste.

The idea of the ecological footprint developed in the early 1990s, as academics started to try to quantify the various demands that humans place on the planet. It aims to help us understand our lifestyles in the context of the Earth's so-called carrying capacity – that is, its total ability to sustainably provide resources and ecological services.

The most comprehensive review of humanity's ecological footprint is the WWF *Living Planet Report*, which is published every two years and tots up the demands made by each country for carbon-absorbing land, crop land, grazing land, forest, fishing ground and built-up land. The 2008 report concluded that if the world continues on its current trajectory then by 2030 we'd need two planets to support us sustainably. If everyone wanted to live as we do in the West, then we'd need more planets still.

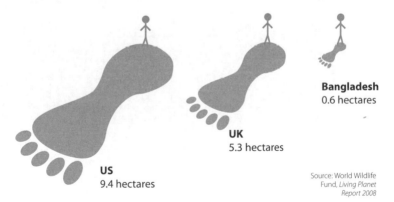

Bangladesh
0.6 hectares

UK
5.3 hectares

US
9.4 hectares

Source: World Wildlife
Fund, *Living Planet
Report 2008*

The diagram above gives some sense of the inequality in the footprints of people around the world. The global average is 2.28 hectares – roughly half of the figure for the typical British citizen and just a quarter of that for an average American.

To read more about ecological footprints, download the latest *Living Planet Report* here:

WWF Living Planet Report panda.org/lpr

Or to calculate your own ecological footprint, and find out how many planets we'd need if everyone lived as you do, visit:

Earth Day Footprint Quiz myfootprint.org

For more information about the ecological footprint system, visit:

Redefining Progress rprogress.org

Do individual actions matter?

The limits of green lifestyles and other ways to get involved

We're regularly told that green-minded individuals can help make a real difference in the fight against climate change. "If we all make small changes we can make a tremendous impact", says one book. "The smallest steps can add up and make the biggest difference", says a local government poster. In a similar vein, the book you're reading right now suggests a whole manner of small changes that you can make to cut back on your carbon footprint in the hope of leading a more sustainable life. But is it actually true that small lifestyle changes can make a big difference?

This may seem an odd question to ask, but it's an important one. And, as this chapter shows, when you zoom out to see the bigger picture there are at least two reasons why making changes to your lifestyle might have a smaller impact than you might expect. That's not to say that trying to live a greener life is pointless. On the contrary, it's critically important that individuals succeed in slashing their own footprints – both to reduce emissions directly and to show government and the rest of society what's possible. At the same time, though, greener lifestyles alone won't solve climate change, and reducing your own footprint is arguably less effective than getting your voice heard politically.

Some bad news about green choices
Carbon caps and energy markets

It would seem reasonable to assume that each time you reduce your carbon footprint – by not driving, not using electricity, or not flying, for example – then you'll be making a commensurate difference to the amount of CO_2 entering the atmosphere. If you reduce your footprint by a tonne, say, then surely you'll be stopping a tonne of CO_2 from being emitted. Unfortunately, the truth isn't always quite so simple, thanks to the effects of carbon markets and the energy markets. Let's take a brief look at each in turn.

How carbon trading can render your greener choices ineffective

In the hope of reducing greenhouse emissions on a large scale, some governments, including the members of the European Union, have signed up to so-called cap-and-trade schemes for carbon dioxide. These schemes work like this. First, a maximum level of CO_2 emissions – a "cap" – is set for a group of polluting industries. Usually the cap will be set slightly below current emissions levels, with a plan to reduce it a bit further each year. Next, permits are printed for each tonne of CO_2 emissions allowed within the cap, and the permits are either given out or auctioned off to the companies in the industries covered. If the companies don't have enough permits for the CO_2 they need to produce in any given year, then they can buy extras from other companies which have more than they need. The idea is that this approach gives every company a direct financial incentive to reduce its emissions, and the market for carbon permits should – in theory, at least – do an efficient job of making sure the cuts are made in the least expensive ways.

Europe's emissions trading scheme – the ETS – sets carbon caps on electricity generation as well as plants producing cement, iron, steel, glass, bricks, ceramics and paper. So far, the scheme has not been particularly successful: the carbon price has sometimes crashed to almost nothing, removing the incentive for companies to be green, and in some cases businesses have been given more permits than they actually need, adding to their profits but doing no good for the environment.

But how does all this affect green lifestyle choices? Surely carbon trading between companies can't change the fact that when you save some energy and emissions, you'll be helping reduce global warming? Actually, it can. The number of permits is decided in advance, so if green lifestyle choices reduce emissions in one part of the economy, the carbon permits will typically be sold to other industries. Let's say you turn off all your electrical appliances for a year, thereby reducing the amount of coal being burned in the local power station. This might avoid a tonne of CO_2 being emitted, but it would also enable the company that owns the power station to sell a one-tonne carbon permit to another business. If this happens, the CO_2 you thought you'd save would still be emitted, albeit from another source.

So is it a total waste of time to reduce your consumption of electricity, paper, glass and other carbon-capped items if you live in a European country? No. If enough people cut their energy use and purchases, then we might collectively be able to make a difference so big that it couldn't be traded away in the carbon markets. And even if that fails, by cutting consumption, green-minded individuals will be making it practically and politically easier for governments to lower the cap in years to come – and lowering the cap *will* make a difference.

Nonetheless, there's no doubt that the current carbon trading system does potentially reduce the benefits of well-meaning green choices. One way around this is to buy and destroy ETS carbon permits equivalent to the emissions savings you make in your own life. That's something you can do surprisingly easily thanks to the charity Sandbag (see p.315).

Of course, much of the typical person's carbon footprint is caused by emissions sources that *aren't* covered by the ETS. Two important examples are vehicle and heating fuels. So if you turn down your gas central heating or put less petrol in your car, then the environmental benefits definitely won't be reduced by emissions trading. Unfortunately, though, they might be partly reduced by the energy markets instead.

How energy savings in one place can be partly cancelled out elsewhere

Schoolbook economics tells us that the price of a commodity is determined by the relationship between supply and demand. If lots of consumers want to buy redcurrants, say, and there are only a small number of redcurrant bushes, then the price will be high. If supply rises (ie more bushes are planted) or demand falls (ie fewer consumers want redcurrants), then the price will drop.

This basic market logic applies to all commodities sold in markets – including the fossil fuels responsible for man-made climate change. The implication, when you stop to think about it, is fairly obvious. If some of us go green, thereby slightly reducing the demand for fossil fuels, but the supply of oil, gas and coal remains constant, then the result will be a drop in the price of fossil fuels. This, in turn, should make it possible for others – individuals, governments and manufacturers alike – to consume more oil, coal and gas than they previously were doing.

Of course, if fossil fuel demand fell sufficiently, then oil, gas and coal extraction would fall too. But economic theory tells us if oil consumption, for instance, is reduced in one home or town or country by a hundred barrels a week, that *won't* equate to a hundred fewer barrels of oil being produced. At least part of the saving will be cancelled out by rising consumption in other homes or towns or countries made possible by the price fall brought about by the reduced demand. In other words, for every unit of energy green consumers manage not to use, a proportion of that energy will be burned elsewhere.

Paradoxically, it might even be the case that – across whole countries – increases in the efficiency of homes, cars and factories can end up leading to *more* emissions. At least that's the prediction of a piece of counterintuitive economic theory known as the Khazzoom-Brookes Postulate. In the late 1980s American economists Daniel Khazzoom and Leonard Brookes tried to show that, since energy efficiency tends to make societies more productive, it increases economic activity as a whole, thereby boosting energy use more broadly. In addition, the efficiency measures make energy cheaper, reducing the incentive for individuals to switch the lights off, turn down the heating and generally be more frugal with energy – the so-called "rebound effect".

Even if we ignore the price of energy, the rebound effect might reduce some of our green choices. That's because if we reduce our use of gas, petrol and electricity, then we'll inevitably free up money that can then be spent on other things. If we spend the extra money on green products and services – such as solar panels or efficient appliances – then the benefits will be doubled. But if, as is more common, we spend it on buying extra everyday goods or holidays, or on buying a bigger home, then our footprints may end up almost as big as before.

A study published in May 2009 by Terry Barker of the Cambridge Centre for Climate Change Mitigation Research estimated that if all the energy efficiency measures recommended by the International Energy Agency were implemented around the world, around fifty percent of the associated emissions cuts would be cancelled out by various kinds of rebound effect by 2030.

It's important not to get too bogged down with all this economic theory. There are plenty of counterexamples that suggest green choices can sometimes have an impact bigger rather than smaller than expected. For example, people installing solar panels don't just generate clean electricity. As they strive to become energy self-sufficient or to make money by exporting electricity to the grid, they tend also to reduce their home energy consumption. (And, of course, after buying the panels they'll typically have less money left to spend on polluting activities.)

Anyhow, as mentioned at the beginning of this chapter, the world needs green-minded individuals to show the rest of society what's possible – and to create markets for new environmentally friendly products and services. Even if half the benefits get cancelled out by other people's actions, the result is better than nothing. Even if *all* the benefits were cancelled out, there would still be a case for making green choices – the simple moral imperative to try and live a sustainable life.

Ultimately, though, there's no way around the fact that some green choices will have a smaller direct impact than we might expect. For that reason, and because climate change is a huge problem that demands a huge response, it's worth aiming to be a green citizen as well as a green consumer. That means getting involved at the community, regional or national level – as described later in this chapter – and articulating your concerns as well as acting on them in your own day-to-day life.

How green are we today?
Taking stock of green and ethical lifestyles

The impact of our green choices – and indeed ethical lifestyles more generally – depends partly on the number of people who get involved. So it makes sense to ask how many like-minded individuals are out there trying to do the right thing for the environment and global society.

Unfortunately, this is something that's very difficult to measure. We can keep tabs on the money spent on specifically green or ethical goods and services, such as organic foods. But it's not so easy to work out what people *aren't* buying or doing on environmental and ethical grounds. It's also difficult to analyse people's motives. For instance, consumers may be choosing energy-efficient fridges to save money rather than to save the planet, or favouring local stores because they're more convenient, rather than because they want to reduce the carbon intensity of the weekly shop.

One way to measure people's environmental concern is simply to ask them, but the results of surveys assessing climate change attitudes range widely. One major study conducted by HSBC in 2007 compared opinions on global warming around the world. In the UK, it found that only 22% of people were seriously worried about climate change, and just 19% were making a significant effort to reduce their carbon footprint. These numbers were less than half of those reported in China and India, and lower than France, the US and every other country surveyed. By contrast, other surveys have claimed to show that a majority of the British population are concerned or very concerned about climate change, with more than a third claiming to have taken steps to reduce their emissions.

Interestingly, despite the fact that climate change looks set to eventually disrupt almost every aspect of human life and almost all the world's ecosystems, many consumers still feel far more strongly about other issues. One report published by the Co-op supermarket chain in 2008 found that only 4% of shoppers put climate change as their leading ethical priority, compared to 21% for animal welfare and 14% for fair trade.

So are consumers as a whole acting on those broader ethical issues, while ignoring carbon emissions? Or are they simply doing nothing? Again, it's hard to say, because claims about the level of consumer concern often clash with the facts observed in the real world. In one poll commissioned jointly by *The Guardian* and Toyota, for example, two-thirds of consumers claimed to make ethical purchasing decisions. But sales of ethical goods and services simply don't bear that out. It does seem to be true, however, that "ethical consumerism" as a whole is on the up. Published annually by the Co-operative Bank, the *Ethical Consumerism Report* keeps tabs on various kinds of green consumer activity in the UK, from recycling and boycotts on certain brands through to people opting for public transport over their car. Since its launch in 1999, the report has charted the steady rise of individuals attempting to be green and fair in their consumer choices, as the graph on the opposite page shows.

Of course, when it comes to climate change, the only truly meaningful barometer of success is the amount of greenhouse gas entering the atmosphere. In the UK, direct CO_2 emissions have stayed fairly level for the last decade; if you also take account of the goods we consume from overseas, then our footprint has risen fairly steeply. By this measure, if green lifestyle choices have achieved anything so far, it's been to fractionally

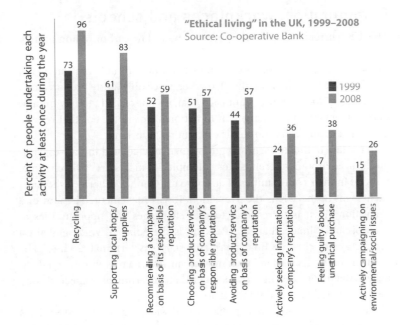

"Ethical living" in the UK, 1999–2008
Source: Co-operative Bank

Percent of people undertaking each activity at least once during the year

1999
2008

Recycling: 73, 96
Supporting local shops/suppliers: 61, 83
Recommending a company on basis of its responsible reputation: 52, 59
Choosing product/service on basis of company's responsible reputation: 51, 57
Avoiding product/service on basis of company's reputation: 44, 57
Actively seeking information on company's reputation: 24, 36
Feeling guilty about unethical purchase: 17, 38
Actively campaigning on environmental/social issues: 15, 26

slow down the rise in the UK's carbon footprint and to boost the profile of environmental issues. The ultimate goal of actually making a noticeable dip in emissions is still a long way off – though of course that's no reason for not continuing to try.

Beyond green lifestyles
Building lower-carbon communities and countries

If greener lifestyles are necessary but not sufficient to solve climate change, what else can we do to make a difference? One option is to scale up our ambitions from the household to the community level. If the people in a village, town or district – or indeed a school or company – work together to reduce emissions, the possibility of making a meaningful cut will be greater, and so will the chance of getting noticed. Another option is to engage with policymakers: either by writing to them directly, or by joining pressure groups that are lobbying for swifter and more meaningful political action on sustainability.

Communities, workplaces and schools

The UK currently has a few purpose-made low-carbon communities, such as the BedZed development in the Borough of Sutton, south London. These are valuable examples that show that low-carbon living can be not only possible but also attractive. However, since only a tiny proportion of the housing stock is newly built each year, it's not realistic to rely on new developments to make a large cut in national emissions. For that reason, greening existing communities is even more important.

In the face of underwhelming government and council initiatives, various local environment groups have attempted to kick-start environmental projects in their communities. The approach that has generated the most excitement in the UK is the Transition Towns movement. The aim of a Transition Town is to respond to the "challenges and opportunities of peak oil and climate change". (Peak oil is the point at which global oil production reaches its final peak and starts its terminal decline. The founders of the Transition Town movement, like many environmentalists, believe that the peak will arrive sooner than governments expect, throwing unprepared economies into chaos.)

The Transition Town concept has spread widely since the first projects were established in Kinsale, Ireland, and Totnes, Devon, in 2005. At the time of writing, more than 150 communities – from Brixton in inner London through to Waiheke Island in New Zealand – are officially listed as Transition Towns, with hundreds more projects in the pipeline. Though few if any of these projects can claim to have demonstrably reduced carbon emissions in their areas by a substantial amount, they have succeeded in raising awareness about climate change, engaging local governments and showcasing a bottom-up thirst for change.

To see whether you live in a Transition Town, or for advice about setting one up, visit this website:

Transition Towns Wiki www.transitiontowns.org

Another way to get involved on a community level is to look into the possibility of setting up a communally owned wind farm to provide electricity as well as a potentially handsome financial return. With community ownership, wind farms can often bypass problems of local planning objections and, thanks to technological and financial economies of scale, investors are likely to achieve much better carbon savings and cash returns than are possible with home-scale green technologies. For more information on community wind, see p.108.

If community-wide involvement doesn't appeal, another way to make a difference is to join forces with colleagues or fellow students to try and reduce the carbon footprint of your workplace, college or school. If there isn't one already, a good first step is to establish an environment group made up of concerned employees or students. The group can meet to discuss and demand environmental improvements – anything from automatic light sensors for toilets to a policy governing the selection of company cars. In many offices, hospitals, schools and other organizations, groups such as these have made a real difference – not least because it's often possible to show a very real financial benefit to reducing energy waste and investing in efficiency.

For advice on what practices and measures you should be asking your workplace or school to implement, a good first point of call is the Start Savings section of the Carbon Trust website, which offers downloadable posters as well as in-depth information.

Carbon Trust www.carbontrust.co.uk/energy/startsaving

Engaging with policymakers

When politicians are asked why their country and the wider world aren't moving faster to tackle climate change, they usually reply that politicians are the slaves of the people and that – as yet – the majority of the electorate is simply not demanding that climate change should be prioritized over other issues. The UK Energy and Climate Change minister Ed Miliband said as much in 2008 when he called for a "popular mobilization" to demand government action on global warming. In private, many other senior politicians say the same.

(Ironically, the day after Miliband pleaded for a climate change movement comparable to the suffragettes or anti-apartheid campaigners, environmental action group Plane Stupid succeeded in shutting down Stansted Airport for half the morning to protest against the government's plans for airport expansion. Ministers lined up to accuse the protesters of selfish and extremist behaviour.)

It doesn't really matter whether politicians are truly searching for a mandate to slash emissions in the short term, or whether they're just using the apathetic electorate as an excuse for their own failures to act. The point is that the more people who demand meaningful government action on climate change, the more likely such action will be to happen. This is pivotal because without a miraculous society-wide environmental awakening, only government action can really solve the problem. A minority of well-meaning people insulating their lofts is a positive step, but a law requiring free insulation to be offered to every home within a year would achieve much more.

Whatever green steps you take in your own life, then, it also makes sense to register your concern with your political representatives. Even a short note to your MP stating that you feel climate change should be a key government priority is better than nothing. For those with more time or commitment, a focused, polite, well-argued letter on a particular point of policy will be more effective still. If you'd like advice on what topics to write letters about, the campaign-group websites listed below can be a useful source of ideas.

The Internet makes it easier than ever to contact your local member of parliament. Just visit the following website and enter your postcode. You'll be presented with information about your MP (including their voting record on climate change and other issues) and given the option of sending them a message.

They Work for You www.theyworkforyou.com

Words can speak louder than actions

As with energy use, it's worth thinking about the bigger picture when making green and ethical shopping decisions. For instance, it's all well and good to favour one high-street brand over another due to their respective environmental or ethical standards. But unless you go to the trouble of *telling* one or both of the companies involved about the issues that influenced your decision, your choice will have less of an effect than it otherwise might have – or it may even have the opposite effect. Most retailers or brands, if sales are down, will assume that the main problem is that consumers are unhappy with current products and prices. Without a comprehensive market-research campaign, they may never realize that a small percentage of the loss is the result of customers going elsewhere because the brand is associated with prolific carbon emissions, toxic chemicals or sweatshop labour. As one commentator put it, the signal the green shoppers are trying to send may become "lost in the general market noise".

For this reason, you may make a bigger difference expressing concerns to a shop's manager as you hand over your credit card than you would by simply not going in. Of course, the ideal approach would be to act *and* to communicate in other ways – making green and ethical choices as well as writing to companies, for example. This takes more dedication than trying simply to be a green consumer, but it's also likely to have a much bigger impact.

Though the really big changes require action from national government, it's also worth engaging with your council. Local authorities control huge numbers of buildings and vehicles, so they often have enormous carbon footprints themselves, in addition to making policy on transport, housing, recycling and other areas that affect the emissions of local residents. One or two British councils have achieved great things on sustainability – most noticeably Woking in Surrey, where the local government has slashed its own footprint and done a lot to raise awareness in the community. Elsewhere, progress has been fairly slow. The author of this book recently interviewed the head of sustainability for one London council; she was unable to state the council's carbon footprint, nor the planned reduction target, nor indeed any other figures relating to the council's environmental policies. Clearly far more could be done by local governments, but in most cases it will only happen with pressure from local residents. To contact your council representatives, visit:

Write to Them www.writetothem.com

Campaign groups and direct action

Another way to try and bring about action on climate change – in addition to greening your lifestyle and contacting politicians – is to join a relevant campaign group. All the major environment membership charities are working hard to push climate up the agenda and they've scored some notable successes – such as the UK's landmark Climate Change Bill, which was brought into law partly as a result of Friends of the Earth's "Big Ask" campaign. Anti-poverty groups such as Oxfam and Christian Aid have also been lobbying and campaigning on climate change, highlighting the threat that global warming poses to the world's poorest communities.

Friends of the Earth www.foe.co.uk
Greenpeace www.greenpeace.org.uk
WWF www.wwf.org.uk
Oxfam www.oxfam.org.uk
Christian Aid www.christianaid.org.uk

There are also many campaign groups focusing on a specific climate change solution. These include Sandbag, which aims to ensure that carbon trading works to reduce emissions rather than simply lining the pockets of polluters. CoolEarth, meanwhile, provides a way to help keep carbon locked up in rainforests that would otherwise face destruction. See p.315 for more on these two charities.

Groups working on climate change and the developing world include SolarAid, which supplies solar lamps (among other items) to poor communities around the world. By substituting lights powered by kerosene, the solar lamps reduce carbon emissions, avoid local air pollution and reduce household expenditure on fuels.

SolarAid www.solar-aid.org

Then there are groups that believe direct action is necessary to bring about emissions reductions. PlaneStupid, for example, campaigns against airport expansion and has gone so far as to invade runways in order to temporarily shut down airports. Activists from Climate Rush, meanwhile, recently glued themselves to statues in Parliament in response to a government announcement on coal-fired power stations.

Plane Stupid www.planestupid.com
Climate Rush www.climaterush.com

Most of the organizations mentioned above are involved in the annual Climate Camp, a protest that sees thousands of people gather to demand action on climate change. The first four took place at Drax, the UK's largest power station; at Heathrow, in response to the proposed new runway; at Kingsnorth, the site of a planned coal power station; and in the city of London, where the focus was on carbon trading.

Climate Camp www.climatecamp.org.uk

The China question
Can the West make a difference anyway?

So far, this chapter has discussed the question of what green lifestyles can achieve compared to changes rolled out by government. But what about the claim that, due to the rapid economic growth in China and other developing countries, there's no point in the UK reducing its emissions even on a national scale? Can a country such as Great Britain, which is responsible for just a couple of percent of global greenhouse gases, make a meaningful difference in the face of inexorably rising emissions from developing giants such as China and India?

A factory producing fibre-optic cables in China. As much as 30% of China's emissions – and half its recent growth in emissions – are the result of producing goods for people in other countries.
Photo: Steve Jurvetson via Creative Commons

China, according to authoritative sources such as the International Energy Agency, overtook the US as the world's biggest emitter of CO_2 some time in 2006. For the last few years, just the *increase* in China's emissions was greater than the total output of the UK. The country is opening new coal-fired power stations every week, and even the economic downturn has done little to buck the country's trend of rising emissions. It's not hard to understand why people find this disheartening, and a reason for not bothering to reduce our own emissions. When you look at the details, however, this argument doesn't really stand up.

For one thing, as we saw in chapter one, China's massive carbon footprint is spread between around 1.3 billion people – almost a fifth of the global population. In terms of emissions per person, China doesn't fare too badly – lower than, say, Algeria or Uzbekistan, and less than half of the figure for the UK. And that's before you consider China's relatively low historical emissions. According to the US government's Energy Information Administration, even if China's emissions continue to soar until 2025, then the total amount of "Chinese" CO_2 in the air will still be half of the total amount of "American" CO_2 at that time

Then there's the fact that, as we saw in chapter two, a large slice of Chinese emissions is caused by manufacturing goods for export. According to recent research by Glen Peters at the Center for International Climate and Environmental Research in Oslo, around half of the recent rise in Chinese emissions can be accounted for by the manufacture of goods that are exported to and consumed in other countries – especially Europe and North America. Arguably China bears some of the responsibility for these emissions, but surely not all.

(Imports and exports are one reason why national boundaries are not the best way to gauge climate responsibility. Another reason is that companies are increasingly international. In 2007, Christian Aid released a report which estimated that the UK's total carbon footprint shoots up from 2% to 15% of global emissions if you include all the overseas activities and investments of companies listed on the London Stock Exchange.)

All told, then, although rocketing Chinese emissions are a very serious concern, China is not necessarily the carbon villain often portrayed. And the country is starting to take steps to curb its use of fossil fuels. In 2004 it introduced fuel-efficiency standards for cars that are stricter than those found in the US, and those standards were tightened further in 2008. This is one example of the policies included in China's "National Climate Change Programme", which aims to increase energy efficiency by 20% in the period 2005–2010. Though this plan doesn't include cutting total emissions, it's a step in the right direction.

As this book goes to press, a senior Chinese politician told *The Guardian* that he expects to see a fifth of Chinese energy being provided by renewable sources by 2020 – a higher proportion than the UK is even aiming for in the same period. It remains to be seen whether this aim will be realized, but it's another important milestone on rising awareness in Asia of the need to slash emissions. In the meantime, there's much that people in developed nations can do to reduce their carbon emissions. Economic growth in China seems a very poor excuse for inaction at home.

Part II

Home

Home energy: the basics

Understanding the footprint of your home

4

When we flick on a light switch or turn up the heating, we rarely stop to think about where our energy comes from and the impacts of its use. But climate change demands that we do just that. Currently fuelled overwhelmingly by gas, coal and oil, domestic heating and electricity account for around a fifth of the total emissions of a country such as the UK. Thankfully, home energy is an area where there's often much we can do to reduce our carbon footprints. According to the Energy Saving Trust, the average British household could reduce its CO_2 emissions by around a quarter – and save up to £340 per year – simply by becoming a bit more energy efficient.

The following two chapters will show you how to make these kinds of efficiency savings. The chapter after that covers higher-level home improvements: renewable heating systems and domestic electricity generation. First, though, let's look briefly at home energy in general.

Assessing your energy profile

If you want to reduce your household emissions, the first step is to understand how your home energy use breaks down. This is an area of some confusion, because the relative environmental impact of heating, cooking, appliances and hot water varies widely in different types of homes. The pie charts overleaf show a typical energy breakdown and total emissions for

Typical home energy use in types of family homes

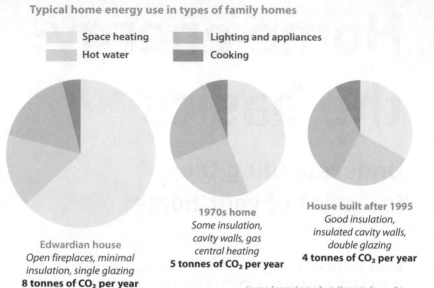

| | Space heating | | Lighting and appliances |
| | Hot water | | Cooking |

Edwardian house
*Open fireplaces, minimal
insulation, single glazing*
8 tonnes of CO_2 per year

1970s home
*Some insulation,
cavity walls, gas
central heating*
5 tonnes of CO_2 per year

House built after 1995
*Good insulation,
insulated cavity walls,
double glazing*
4 tonnes of CO_2 per year

Source: Energy Saving Trust, Domestic Energy Primer

three different types of house. As the charts make clear, space and water heating are the biggest consumers of domestic energy overall – especially in older, leakier homes. This flies in the face of received wisdom about the energy importance of gadgets and phone chargers, but it makes sense of a basic fact of physics: it takes a lot of energy to raise the temperature of things.

Energy versus emissions

When it comes to climate change impacts, the pie charts shown above are a little misleading. That's because lighting and appliances are typically powered by electricity, whereas heating is most commonly gas-fuelled. As the box on the opposite page explains, electricity from the national grid is currently a far more carbon-intensive energy source than gas. When you focus on emissions, then, the picture is rather different: the lighting and appliances in the leaky Edwardian home cause around a third of the emissions; in the modern, well-insulated home they cause more than half.

For homes off the gas network, the energy and emissions breakdown varies widely depending on whether the home is heated by wood (which is close to carbon neutral), oil (282 grams of CO_2 per unit of heat), coal (313 grams) or electricity (a massive 537 grams).

Of course, the exact figures vary from home to home, so the ideal way to understand your energy use and emissions is to have an expert assess your property. Home energy audits can usually be arranged for free via your local Energy Saving Trust centre. Call 0800 512 012 for more details.

Why gas is greener than electricity

Electricity may be a more versatile energy source, but when it comes to creating heat – whether for radiators, water or cooking – gas is currently a far more climate-friendly option. According to the standard "conversion factors" published by the government, a unit of gas creates 206 grams of CO_2, whereas a unit of grid electricity currently causes 537 grams. Electricity emissions are due to fall over time as wind turbines and other renewable technologies start to take over from fossil fuel plants, but for the foreseeable future, gas will remain much greener.

The benefits of gas are obvious when you think about it. With a gas boiler or cooker, you take a fuel and burn it to create heat exactly where you want it, exploiting the fuel's full energy potential. With an electric heater or cooker, by contrast, someone else takes the gas – or, even worse, coal – and burns it in a power station. At least half the energy in the fuel is wasted as heat and vented out of the power plant's chimney; what remains is turned into electricity, some of which is then lost in transmission via the national grid en route to your home; what remains at this point slowly warms your radiator or hob.

Greening your home: where to start

If you live in a home constructed in the last ten years, or a home that's already had an eco-makeover, then your heating system and insulation are probably already fairly green, and you should focus first and foremost on appliances and lighting (chapter six). You might also want to consider the possibility of installing a renewable energy system (chapter seven).

If, however, you live in an older house, then heating and hot water (chapter six) are the most pivotal areas. More specifically, it make sense to start with insulation. There are three main reasons for this. First, adding insulation is generally a fairly inexpensive way to achieve substantial carbon savings, as it lets you burn substantially less gas, electricity or oil to heat your home. Secondly, the benefits of insulation will typically last as long as the home, so you're locking in large savings for a long period.

Thirdly, and less obviously, insulation somewhat reduces the importance of the inefficiency of your appliances and lighting. Inefficient electrical appliances are inefficient precisely because a large slice of the energy they consume is lost as heat. In an extremely well-insulated home, this "waste" heat won't actually be wasted – at least not in the winter when you have the heating switched on. Indeed, perfectly insulated homes – such as the *passivhaus* buildings first developed in Germany – don't have a central heating system at all, because the warmth generated by humans, cooking and appliances, perhaps with one small top-up heater, are sufficient to keep the building warm even in winter.

This isn't to say that you should worry about insulation and nothing else. As we've seen, electricity is currently a far more polluting source of energy than gas, so setting out to heat your home with the waste heat of inefficient electrical appliances wouldn't be exactly sensible. Moreover, you won't want or need the extra heat from the appliances in the summer. Nonetheless, if you live in a leaky house it does make sense to focus first on insulation and heating.

Grants and offers

Various grants are available to people wanting to make their homes greener and more energy-efficient, and there are likely to be more appearing in the next few years. In 2009, the Conservative party issued a paper promising that if they form the next government they will offer £6500 to each household for eco-renovations, repayable via utility bills. At the time of writing, there is nothing quite so good as that, but there are a couple of schemes worth exploring.

Warm Front

This government scheme offers up to £2700 – or £4000 if oil central heating is involved – for insulation improvements. You'll need to be either a pensioner or on a low income to qualify.

Warm Front www.warmfront.co.uk • 0800 316 2805

Low Carbon Buildings Programme

The Low Carbon Buildings Programme is a government scheme offering grants for the micro renewable technologies discussed in chapter seven. A successful application depends on your home already having basic levels of energy efficiency in place, including 270mm of loft insulation, cavity wall insulation if possible, low-energy light bulbs, thermostatic radiator valves (TRVs), and a room thermostat and programmer. The application process is fairly simple and you'll normally get a decision within thirty days. For more information, see:

LCBP www.lowcarbonbuildings.org.uk • 0800 915 0990

The Low Carbon Buildings Programme may be phased out once the government unveils a feed-in tariff in 2010 (see p.104).

Utility companies and local authorities

Under the government's Carbon Emissions Reduction Target, utility companies are obliged to spend a slice of their income on projects designed to reduce domestic energy use – by distributing low-energy bulbs, for example, or subsidizing insulation or energy monitors. Read the literature that comes with your bill to discover deals and freebies, or scan the company's website.

Also try your local authority. Some councils – such as Kirklees in West Yorkshire – have started to implement impressive schemes to distribute free insulation to all homes that need it.

Selling an energy-efficient home

When deciding whether to invest in energy-efficiency and renewable-energy measures for your home, bear in mind the benefits they might bring if you ever sell the property. Since 2007, everyone selling a home has had to produce a Home Information Pack. As part of this process an inspector produces an Energy Performance Certificate, which ranks the home from A to G for efficiency and provides recommendations for improvements.

A poll carried out by the Energy Saving Trust suggests that 70% of homebuyers see energy efficiency as an important feature. Only 21% consider improving this aspect of their home before putting it on the market, even though 64% of the same people claim they would avoid buying houses with old boilers, single glazing and insufficient insulation.

Energy Saving Trust www.energysavingtrust.org.uk • 0845 727 7200

Moving on: is it bad to buy an inefficient home?

Moving home provides a clear opportunity to reduce your carbon footprint. If you choose a house or flat that's highly insulated or comes with a renewable energy system, then your gas and electricity emissions could be substantially reduced for as long as you stay there. Of course, few of us would be prepared to buy a home solely on the grounds of energy efficiency. Other factors matter too, such as aesthetics, and unfortunately for those green-minded homebuyers who have a taste for period properties, the older the house, the less efficient it tends to be. So is it a climate crime to buy a beautiful but leaky old country cottage or Victorian semi?

This is one case where it pays to see the bigger picture. Although buying (or indeed renting) an inefficient property might make it more difficult to get your own carbon footprint down to a sustainable level, the overall impact should be neutral – or even positive. After all, unless all period properties are pulled down, then *someone* will end up living in them. And emissions will probably be lowered overall if the people most concerned about climate change are occupying the least efficient homes – and vice versa. Moreover, if you move into an inefficient building and give it a green-makeover, using the techniques described in the following chapters, then you'll be making the biggest difference possible.

Overall, then, there's little environmental benefit to choosing your home based on its efficiency. A more important consideration is to try and buy a place that will require you to drive as little as possible – for example, a home fairly near to your work, family and friends.

Understanding energy units

Energy consumption is defined in terms of watts and hours. If a 100 watt light bulb is used for ten hours, then the power used is 1000 watt hours. For ease we refer to 1000 watt hours as 1 kilowatt hour (kWh) and this is the standard unit that appears on our electricity bills. Gas bills normally specify the number of cubic metres of gas consumed but also convert the figure into kWh.

At the time of writing, typical prices (including standing charges) are 4p per kWh for gas and 14p per kWh for electricity. Each year, an average three-bedroom semi-detached house uses around 15,000kWh for space heating, 5000kWh for water heating and 3000kWh for lighting and appliances, costing a total of around £1000. This results in around six tonnes of CO_2 emissions – or even more if you factor in the extraction, processing and delivery of the fuels to the home or power station.

Heating & hot water

How to warm your home without warming the planet

Domestic space and water heating accounts for at least a tenth of UK greenhouse emissions, and as much as half of the emissions that each of us has direct control over. There are plenty of ways to reduce these emissions – and some of these changes offer substantial financial savings. The first step is to minimize waste by tweaking the way you use your heating. The second is to make sure your home is well insulated. The third step is to consider upgrading to a more efficient boiler or investing in a renewable system powered by wood, sunlight or the ground.

Step 1: Tweaking your heating
Small steps to increase efficiency

Most households can achieve significant cuts in their CO_2 emissions just by making a few adjustments to the way they use their existing heating setup. Here are some good ways to get started.

▶ **Turn it down** Reducing your room temperatures by just a small amount can make a disproportionate difference to your energy consumption. You may find you sleep better, too. For rooms, try 18°C and wear an extra layer. Each degree you lose will save around 100–250kg of CO_2 emissions (and £20–50 on your bills) each year

▶ **Avoid over-hot water** If the water coming out of your taps is scalding, then you're probably wasting energy. Set your boiler or tank to deliver the water at 60°C. (Avoid going below this level, though, or your tank could start harbouring the bacterium that causes Legionnaires' disease.)

▶ **Use the controls properly** Be sure to set your heating programmer carefully so that it matches your daily or weekly routine. (If you don't have a programmer, install one, as described on p.64.) Remember to adjust the setting as your routine changes, and to turn the heating down or off half an hour or so before you go out.

▶ **Don't heat empty spaces** If you have thermostats on individual radiators, turn them down or off in rooms which are rarely used. If you don't have radiator thermostats, consider adding them (see p.64).

▶ **Get your system serviced** A well-maintained boiler is likely to be more efficient – and to have a longer life – so don't let too many years pass without having yours serviced.

▶ **Bleed your radiators** At least once a year, use a radiator key to vent any trapped air from each radiator in your home.

▶ **Heat rooms not walls** Foil reflectors behind radiators can reflect heat back into the room. This helps a room feel warm more quickly, and – for radiators on external walls – can reduce heat loss. It's possible to make your own reflectors, but you may get better results buying them off the shelf. They're inexpensive and available online and in hardware shops.

▶ **Close the curtains** Draw the curtains at dusk, before the warmth starts to escape. Also consider adding linings to curtains to increase their insulating power. These can be purchased inexpensively and have the bonus of reducing the amount of noise and light the curtains let through.

▶ **Bust draughts** Inexpensive old-fashioned draught busters – such as "snakes" for the bottom of doors – can make a room warmer and more comfortable.

▶ **Keep your tank warm** If you have a hot water tank that doesn't feature a modern insulating coating, be sure to equip it with a high-quality insulating jacket. This could reduce the energy required to heat your water by as much as a quarter. Alternatively, consider upgrading to a combi boiler that heats only the water you actually need (see p.64).

Step 2: Insulation
How to reduce heat loss

It may not be as glamorous as installing a solar panel, but improving insulation can have just as big an effect. Indeed, in terms of cost-effective ways to reduce greenhouse-gas emissions, making our homes less thermally leaky sits towards the top of the list, alongside cutting back on air travel. Decent loft insulation alone can reduce the annual CO_2 footprint of a typical UK home by a tonne per year and reduce heating bills by a fifth. Wall insulation, though more expensive, can often have an even bigger effect.

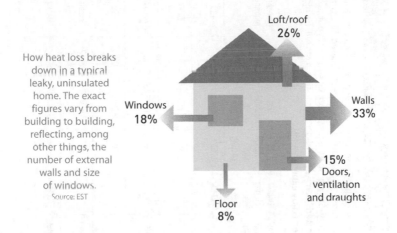

How heat loss breaks down in a typical leaky, uninsulated home. The exact figures vary from building to building, reflecting, among other things, the number of external walls and size of windows.
Source: EST

Loft/roof 26%

Walls 33%

Windows 18%

15% Doors, ventilation and draughts

Floor 8%

Loft insulation

Current building regulations state that loft insulation in new buildings should be 20cm. If your current insulation is 10cm or less, you should seriously consider topping it up to the 20cm level, and possibly adding even more. If you want a grant to install a renewable-energy technology, you'll need at least 27cm.

It's well worth exploring the various grants available for loft insulation (see p.52). If you can afford it, however, you may want to forgo any grants and opt for the greenest possible insulation materials. As a rule, standard mineral-based materials use more energy and chemicals in their production – and are less likely to be locally sourced – than those that use natural materials such as wool or flax. The latter, though often more

Insulation options: costs, savings & payback periods

The exact cost and payback period for each type of insulation and draught-proofing depends on your specific home and heating system. The following figures from the Energy Saving Trust are based on a typical three-bedroom, semi-detached house. If your home is particularly leaky, or heated by expensive oil or electricity, then the savings will add up more quickly; if you live in a well-insulated flat, with few exterior walls, the savings will be lower. Of course, if your home is bigger or smaller than average, the installation costs will be higher or lower accordingly.

	Cavity wall insulation	Internal wall insulation	External wall insulation	Double glazing	Loft insulation (0–270mm)	Loft insulation (50–270mm)	Floor insulation	Draught-proofing	Filling gaps between floor and skirting board	Hot water tank jacket	Primary pipe work insulation
Approximate annual saving on bills	£160	£470	£500	£140	£205	£60	£50	£30	£25	£40	Around £10
Approximate cost, using a professional	£250	From £42/m²	£5600		£250	£250		£200			
Approximate payback time	2 years		11 years		1 year	4 years		7 years			
Approximate DIY cost					£300	£200	£90	£90	£20	£12	£10
Approximate DIY payback time					From 2 years	4 years	From 2 years	3 years	1 year	5 months	Up to 1 year
Approximate annual CO₂ saving	800kg	2.4 tonnes	2.5 tonnes	720kg	1 tonne	300kg	250 kg	150 kg	130 kg	195 kg	65 kg

expensive, allow for better circulation of air and help to avoid the retention of toxins in a building, linked by some to "sick building syndrome". Options include:

Thermafleece www.secondnatureuk.com • 01768 486 285

Made from the wool of British sheep, Thermafleece is less intensively produced than glass fibre and, according to its producers, pays back the carbon footprint of its manufacture seven times faster. (If you considered the methane emissions from the sheep, however, this benefit might be greatly reduced.) It can be installed without gloves or protective clothing. For a typical 40m² loft with 10cm of glass-fibre insulation, it would cost around £700 to add a further 10cm of Thermafleece, which would pay for itself after approximately a decade.

Warmcel www.excelfibre.com • 01685 845 200

Warmcel is made from recycled newspaper, with inorganic salts added to provide fire resistance. The material is blown into place with a hose (ideal for difficult-to-access spaces) so there are no offcuts or different roll thicknesses to worry about. Warmcel is more expensive than glass fibre but cheaper than Thermafleece, costing £8 per square metre to upgrade from 10 to 20cm. A 40m² loft would therefore cost around £300, paying for itself in five years or so.

Wall insulation

Most houses built in or after the 1930s have cavity walls – an inner and outer wall with a gap in between. You can usually spot cavity walls as they're relatively thick: around 30cm, compared to around 23cm for a typical solid wall. The pattern of the brickwork can also be a useful clue. With cavity walls, each brick is equally wide, whereas with solid walls many bricks will be placed sideways on and appear half as wide.

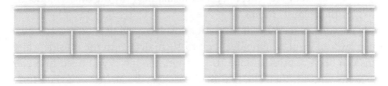

Typical brick patterns of cavity walls (left) and solid walls (right)

If you've got cavity walls

If your home has cavity walls, then making sure the cavity is full of insulating material is a no-brainer. Filling an empty cavity can lead to huge energy savings for a fairly small initial outlay, and it's a quick job that causes minimal disruption. Holes are drilled in the building's exterior wall and insulation material, which can be foam, mineral wool or some other option, depending on your budget, is injected into place. If you're not sure whether your cavity walls have already been insulated, then look for small circular marks left by the drilling process. Alternatively, ask a local insulation service to come and take a look. Note that an estimated 1.75 million homes in the UK have cavity walls that are unsuitable for filling. If yours if one of them, consider solid wall insulation instead.

If you've got solid walls

Around a third of the UK's 24 million dwellings were built before 1930 and have solid rather than cavity walls. It's generally more costly and labour intensive to insulate these kinds of walls, though it's perfectly possible and may well be the single most important step you can take to reduce your home's footprint.

The main decision is whether to go for external or internal insulation. The internal option involves adding insulating material to the inside face of the exterior walls – either boards within a narrow wooden frame or a flexible material known as Sempatap. The latter is rolled on almost like wallpaper, can be decorated with any finish (emulsion, wallpaper or even tiles) and has the benefit of combating condensation and black mould.

Internal wall insulation will of course reduce your room sizes, though if you use Sempatap you'll lose only 1cm along the length of each affected wall (usually one or two walls per room). The typical cost will be around £45 per square metre, plus the price of redecoration. At this price it might take more than a decade to repay the initial outlay. However, DIY enthusiasts can save money by doing the installing themselves, and it's also possible to cut the upfront cost by insulating only the most frequently used (or the coldest) rooms.

By contrast, external wall insulation involves paying someone to add a thick layer of insulating render – up to 10cm deep – to the outside walls. This can be very pricy if done in isolation, though if you need to repair the walls anyway, then the marginal cost may be as low as £1800 for a medium-sized home, in which case it might pay for itself in just five to six years. Of course, exterior insulation will drastically change the look of your building. In conservation areas this may rule it out completely. Elsewhere, you'll need to balance the energy savings with aesthetic concerns. The final finish can be flat (wet render), pebble-dashed or, for an extra cost, clad in wood, brick slip, clay or aluminium.

External insulation will typically produce slightly better energy savings than the internal option – though the difference is fractional.

Double glazing

Old single-glazed windows are a major source of heat loss, so it's worth considering double – or even triple – glazing. However, new windows are expensive, and the payback period can be as much as twenty years. For that reason, if you have a limited budget and your single-glazed windows don't feel particularly leaky, then it's probably sensible to start off by adding loft and wall insulation. On the other hand, if you have large or numerous draughty windows, the payback period could be much shorter.

If money is no object, the greenest choice for double- or triple-glazed windows are those with wooden frames made from timber from certified sustainable sources. Frames made from PVC don't last as long and can't be recycled, while metal can be problematic in terms of thermal conductivity. Whatever type of frames you opt for, favour windows with a good energy-efficiency rating. This is easily done where BFRC labels (like the one pictured) are on display. Unfortunately, retailers aren't legally obliged to show these labels, but you can see how each window brand compares on the BFRC website:

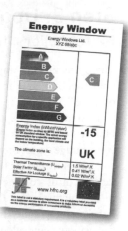

BRFC www.bfrc.org

If you have a more modest budget, you might want to explore DIY secondary glazing, which involves adding extra (openable) windows either inside or outside your existing ones. These aren't particularly neat but they can work well.

The very cheapest method of secondary glazing is a special cling-film-like material applied to standard window frames. You can do a whole house for less than £20, but it's a fiddly, time-consuming job, needs to be repeated each year and will produce only small energy benefits.

Floor insulation

People rarely think about insulating their floors, but some homes lose as much heat to the ground as they do through the windows. For houses with wooden floors on the ground level, the insulation process is fairly simple. Some or all of the boards are temporarily taken up and the insulation is either rolled or blown between the joists. For homes with solid floors, the process is a little more involved. Adding the insulation is easy enough (sheets of material are placed above the underlying concrete, with chipboard and a finishing layer put on top) but the floor will be raised slightly so you'll need to shorten doors and raise the skirting boards. On the plus side, adding insulation over a concrete floor can rapidly increase the speed with which rooms warm up after the heating is switched on.

When insulating floors, it's particularly important to focus on the area where the floor meets the exterior walls, as this is where much of the heat loss occurs. For the same reason, it makes good sense to fill the gaps between the floors and the skirting board, even if you decide not to insulate underneath.

Step 3: Upgrading your system
Better boilers, heating controls and switching fuels

There are various ways to give your heating system a green overhaul. Which is best for your home will depend on a number of factors – including your budget and the fuel that you currently use for heating. By far the most common heating fuel is natural gas. Consisting largely of methane, this gas is the vapour equivalent of coal and oil, formed underground by decomposing matter over millions of years. Gas extraction,

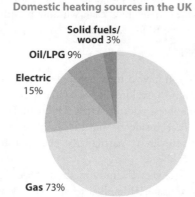

Domestic heating sources in the UK

Solid fuels/
wood 3%

Oil/LPG 9%

Electric
15%

Gas 73%

processing and burning are major sources of carbon dioxide but as we've seen this fuel is both cheaper and less harmful to the environment – in terms both of greenhouse gases and local air pollutants – than either oil, coal or grid electricity. For this reason, renewable heating systems such as ground-source heat pumps and wood-fired boilers are particularly tempting, both financially and environmentally, for those 27% of UK households that aren't connected to the gas network. These alternative systems are described in the next chapter. First, let's look at options for increasing the efficiency of a regular boiler and heating controls.

Electric heating

Compared to gas, electricity produces more than twice the CO_2 per unit of heat. There are currently around 1.6 million UK residences heated by electricity – around 15% of housing stock. Many of these are rural houses that aren't connected to the gas network, but even in cities a significant number of homes – especially rented flats – use electric radiators and storage heaters. It's strange that whilst the government has tightened up energy conservation measures in building regulations they still haven't banned electric heaters in gas-connected areas. As a consequence many well-insulated residences built in the 1990s produce as much CO_2 via heating as poorly insulated houses from the 1890s.

If you own a home on the gas network which uses electric heating, then seriously consider switching over. If you're off the gas network, explore the alternative heating systems described in chapter seven. If you're renting an electrically heated house or flat, probably the best you can do is use your heaters as efficiently as possible. With simple electric radiators, this may mean nothing more than setting thermostats at sensible levels and switching the heaters off a while before you go out. With storage heaters, which build up heat reserves in the night, when electricity is cheaper (and less carbon intensive), it's a bit more complex. Such heaters usually have two controls: an input and an output. The input regulates the amount of heat that is stored in the heater at night; turn this up during cold weather and down during warm weather. The output dial allows you to control when you release the stored heat. Turn this down a while before going to bed or leaving the house.

Upgrading your boiler

If you want to increase the efficiency of your heating system rather than go all out with a renewable setup, then consider upgrading your boiler. If your current model is more than ten years old, upgrading to a modern condensing model could easily slash your heating bills and emissions by a third. Condensing boilers extract more usable heat from each unit of fuel thanks to a heat exchanger that captures the energy in the hot gases exiting the flue. They can extract more than 90% of the energy in the fuel, compared to around 80% for a new non-condensing boiler and just 60–70% for an older boiler.

You can check the efficiency of your current boiler by looking it up on www.sedbuk.com. Or to quickly see whether it's a condensing model, look outside at the flue. If it lets off steam when the boiler is in use, then it's probably a condensing model; if it doesn't, then it's probably a non-condensing model and an upgrade will be well worth considering.

Combination boilers versus hot water tanks

If you currently have a hot water tank, then upgrading your boiler is a good opportunity to switch to a combination – or "combi" – boiler. These save energy by heating water on demand rather than filling a tank with hot water that may or may not be used. As a bonus, you'll reclaim the cupboard space currently occupied by your hot water tank, and you'll never again run out of hot water. (On the flip side, losing your tank will mean you can't opt for a solar water pre-heat system; see opposite page.)

A high-efficiency condensing combi boiler costs in the region of £750–1000, plus the price of having it installed, but they can pay for themselves in as little as four years in some large houses; between five and eight years would be more realistic in most homes.

When shopping for a boiler, be sure to choose one bearing the Energy Saving Recommended label. For a list of accredited models, see:

Energy Saving Trust energysavingtrust.org.uk/Energy-saving-products

Also think carefully about what size of boiler to opt for. Some home energy experts bemoan the fact that homeowners often install more powerful boilers than they actually need, resulting in slightly less efficient operation. Perhaps the best approach is to ask a few installers for their view and to go with the smallest model recommended.

Better heating controls

Whether or not you're buying a new boiler, it's worth considering a new set of heating controls. Modern controls have more accurate thermostats, control boilers more intelligently and offer more flexible temperature programming. Used properly, a good set of controls could slash your heating bills by up to a quarter in return for an initial outlay of just £150–300.

Also consider installing thermostatic radiator valves (TRVs). For little more than £10 each, these allow you to control each room's temperature separately. TRVs can make a fairly significant difference in bigger houses, but they shouldn't be put in rooms where there are already wall-mounted thermostats, or the system will get confused trying to respond to both.

Thermostatic radiator valves allow you to set different temperatures in different rooms

How to make an efficient system even greener

Pre-heat systems

A pre-heat system warms the otherwise cold water coming into the boiler. This reduces the amount of fuel the boiler needs to burn to heat the water to the desired level and can also reduce the delay that you experience while waiting for hot water to reach the taps. There are two types of pre-heat system:

▶ **Solar water pre-heat** This option uses a standard solar hot-water system (see p.96) to warm the water before it's fed into the combi boiler. It's quite an expensive setup to install and you'll need a tank as well as a boiler. Also, note that it will only work with certain boilers, so be sure to check with your boiler manufacturer before ordering.

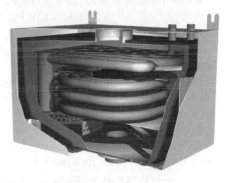

▶ **Zenex Gas Saver** This is a more compact solution – effectively a "top box" that bolts on to a combi boiler. It extracts otherwise wasted energy from gases exiting the flue, using it to pre-heat the incoming cold water feed, and makes the boiler operate more efficiently. The box costs £575 (plus installation costs) but it can reduce gas use by around a fifth. The Gas Saver can currently be used with all Alpha and Veissmann boilers and many others besides. For more information, see:

Zenex Gas Saver www.zenexenergy.com • 0800 328 7533

Lagging the pipes

Depending on where your boiler or hot water tank is situated in relation to your taps, it can take up to a minute to actually get any warm water out. In this time you will waste up to twelve litres of water and a whole lot of gas or oil. Moreover, once you turn off the tap, warm water remains in the pipe. If you need hot water again soon afterwards, this water will still be warm, but if it's left for an hour or so then its thermal energy will usually have dissipated and you'll have to wait another minute for hot water. For this reason it pays to insulate the pipes that take the water from your boiler to your taps. This is best done when the pipes are installed, as some of them will be inaccessible once the system is complete. But even for existing heating set-ups it's possible to reduce thermal waste by adding inexpensive lagging to exposed hot-water pipes in lofts, cupboards and other accessible areas.

Air conditioning
Alternatives to electric AC units

Once relatively rare in the UK, air conditioning is becoming increasingly popular. This trend has been driven by the heat waves made more common by climate change. That's somewhat ironic, since AC units themselves are quickly becoming a significant contributor to global warming. One recent report estimated that by 2020 domestic air conditioners in the UK could release almost five million tonnes of CO_2 per year.

A typical home AC unit running at full power uses around 1–2 kilowatts of electricity, which makes its carbon footprint roughly equivalent to that of a fan heater. Greener ways to keep interior temperatures down include fans – which are far less energy-hungry – and adding pale, reflective blinds to south-facing windows. Good insulation also helps, as do efficient bulbs and appliances. It makes absolutely no sense to waste money and energy on warming your home against your wishes with old-fashioned light bulbs and appliances only to use around twice the energy again to get rid of that unwanted heat.

A more drastic solution advocated by some environmentalists is to paint your roof or exterior walls white. This should help keep the home cool in the summer as well as bouncing some sunlight back into space, thereby reducing global warming directly as well as indirectly. Perhaps the most elegant way to keep a house cool, however, is to employ the old-fashioned technique of planting large trees to the south aspect. Once mature, these trees will block some of the sunlight hitting the home in the summer, but not in the winter, when the branches are bare and the extra heat will be welcome.

If you're really serious about cooling, one final option is to consider a ground-source pump capable of air conditioning as well as heating (see p.98). Or if you must use an electric A/C unit, at least try to avoid pushing the temperature dial below 25°C.

Electricity & appliances

From green gadgets to green tariffs

There can be no feasible solution to climate change that doesn't include a reduction in the carbon dioxide emitted from power stations. Most countries, including the UK, generate the overwhelming majority of their electricity in power plants fired by coal or gas. These plants kick out approximately a quarter of the greenhouse gases released by humans each year. Anyone wanting to reduce the emissions of their electricity use has three approaches at their disposal. This chapter covers two of them: reducing electricity use and switching to a greener tariff or supplier. The next chapter covers the third: installing solar, wind or micro hydro equipment to generate electricity at home.

Perhaps surprisingly, there's a case to be made that none of these three steps will have much impact. As we'll see, green tariffs don't typically deliver what most consumers would expect. And, as discussed in chapter three, the European Emissions Trading Scheme may diminish the benefits of reducing your power consumption or generating your own electricity. For this reason, you may want to consider joining Sandbag and removing carbon permits from the market in addition to reducing your own energy use directly (see p.315).

The big picture

The current "fuel mix" of the UK electricity grid is dominated by natural gas, coal and nuclear energy, which represent 44%, 33% and 16% of the total respectively. The renewable contribution is less than 6% and much

of this comes not from familiar clean energy technologies such as wind or solar but from burning the gas formed as our rubbish decomposes in landfill sites.

There are, of course, targets for increasing the contribution of renewable power sources, including a long-standing government aim to generate 10% of the UK's electricity from renewables by 2010. This target now looks impossible to hit, so the focus has shifted to the European targets for 2020. By that year, the UK has a legally binding obligation to generate 15% of *all* its energy – including transport and heating fuels as well as electricity – from renewable sources. Meeting this target would probably necessitate at least a third of electricity to come from renewables.

Various technologies could help the UK meet the 2020 target, including tidal and wave power and solar electricity. But wind power is seen by most experts as the single most important clean energy source. Unfortunately, the erection of wind turbines continues to be consistently

The dirty power behind our plug sockets: a bellowing smokestack at a fossil-fuel power station

held up by opposition to planning applications.

So what can individuals do about all of this? One approach is to install some renewable generation capacity in your own home (see chapter eight). Another is to voice your support for wind farms and other large-scale renewable power projects both in your local area and elsewhere. That might mean contacting your political representatives or it might mean getting involved in local pro-renewables campaigns. For advice on how to help persuade the local planning authorities to give wind farms the green light in your area, visit:

Yes2Wind www.yes2wind.com

Using less power
How to reduce your electricity use

Reducing household electricity use isn't very difficult and certainly doesn't have to mean drastic lifestyle changes. At present a large proportion of our household power is simply frittered away by the likes of energy-inefficient fridges, overfilled kettles and TVs on standby. As this chapter shows, we can take a decent slice out of our electricity consumption simply by reducing this obvious waste and by replacing some old appliances with newer more efficient models. With a little more commitment and an energy monitor, a fairly radical cut is often possible.

British householders currently spend around £400 per year each on domestic lighting and appliances, so there's a good financial incentive, not just an environmental one, to cut back. A good first step is to take a look at the chart below, which shows a breakdown of electricity use in the typical home. Note that this chart excludes heating and hot water, which are most commonly provided by gas or oil. If your home does have electric radiators and water heating then these will probably constitute the vast majority of your electricity bill – perhaps as much as 75% – and it would be well worth considering an alternative heating system (see p.95).

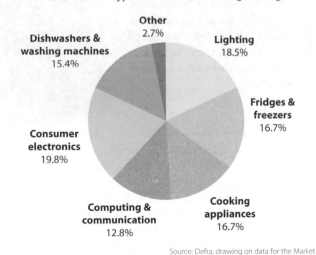

Power use in a typical UK home, excluding heating

Other 2.7%

Dishwashers & washing machines 15.4%

Lighting 18.5%

Fridges & freezers 16.7%

Consumer electronics 19.8%

Computing & communication 12.8%

Cooking appliances 16.7%

Source: Defra, drawing on data for the Market Transformation Programme

Kitchen appliances

In many homes the kitchen is by far the most energy-intensive room, containing, among other items, a fridge, freezer, dishwasher, cooker, kettle and microwave. Let's take a look at each of these in turn.

Fridges & freezers

Household fridges and freezers run nonstop and account for around a sixth of the typical home's electricity bill. So if you have an inefficient model, it's well worth considering an upgrade. Compared to a typical ten-year-old fridge, an efficient new one could pay for itself in just a few years as well as making CO_2 savings from the moment you plug it in. When shopping for a fridge or freezer, always look out for one marked A+ or preferably A++ for energy efficiency, and opt for the smallest model that will comfortably meet your needs. The efficiency ratings are based on the energy consumption per unit of storage capacity, so a large fridge may have the same rating as a smaller one but actually consume far more energy.

If you don't want to invest in a new fridge you could still make energy savings with a SavaPlug, available from various websites and shops. It replaces the fridge's normal plug and has a sensor that reduces the amount of energy needed to pump the refrigerant around the fridge. Savings of more than 20% can be achieved, but before buying be sure to check that your model isn't on the incompatible list at:

Savawatt www.savawatt.com

Whatever type of fridge or freezer you have, its energy consumption is influenced by the amount of time the door is left open and – less obviously – by how clean and ice-free it is. So defrost regularly and once in a while check the grille at the back for dust and dirt. This will reduce power use and lengthen the fridge's working life.

One important but often overlooked consideration is the placement of your fridge and freezer. Cooling appliances have to work much harder if the temperature surrounding them is high, so it's important to always place them away from sources of heat, such as cookers, boilers, hot water pipes or sunny windows.

When's the right time to upgrade appliances?

The standard green advice of opting for the most energy-efficient appliances available is fine when your old model dies and needs to be replaced. But what about old appliances that still function? How bad do they have to be before it's worth binning them and replacing them with new models? In other words, how do you compare the emissions saved by an efficient machine with the emissions caused by disposing of your existing appliance and manufacturing and delivering a replacement?

Comparing the energy consumption of an old and new machine is perfectly possible with an energy monitor (see p.87). If you don't have one, as a rough rule of thumb, upgrading from a fairly typical machine, five to ten years old, to a new one rated best-in-class for efficiency will typically reduce energy consumption by around 65% for fridges and freezers, 40% for dishwashers and 33% for washing machines. Unfortunately, however, there are few comparable figures to help you work out the energy and emissions used in the production and transportation of the new appliances. Even if this data was available, there would be no way to tell how long your new, greener appliances were going to last, which would make a meaningful calculation difficult.

As ever, then, it comes down to common sense and broad-brush calculations. If you have an old or leaky fridge or freezer, then it would almost certainly be sensible to upgrade to a model rated A+ for efficiency. But if you have a middle-aged dishwasher that you only use once a week, then you might be better sticking with the old model.

When it comes to old-fashioned light bulbs, it's always a good idea to replace them straight away. The energy used in producing low-energy bulbs is only a tiny fraction of the energy they save, so every extra hour you use the old bulb will cause a needless waste of electricity.

Dishwashers

As with washing machines, the vast majority of the energy consumed by a dishwasher is used to generate heat – both to warm the water and, for drying cycles, to warm the air inside the machine. So if you do use a dishwasher, always select the coolest setting that will get the job done.

But should you use a dishwasher in the first place? How do they compare with washing up by hand? This depends on the individual machine, the efficiency of your hot-water heater and, most importantly, how economical you are when washing up by hand. One much-cited 2004 study from the University of Bonn suggested that dishwashers use less than half the energy of the typical person at a sink. However, the study assumed the dishwasher is used efficiently (with full loads and no pre-rinse) while giving figures for hand-washing that can be slashed with just a bit of care. For example, soaking washing in a bowl instead of leaving it to stand, limiting the dregs that get mixed with the soapy water when washing up, and using cold rather than hot water for rinsing can all substantially reduce the amount of energy required to wash up by hand. If you follow these steps, and you have a fairly efficient boiler, then you can fairly easily match a modern dishwasher for energy use.

But why bother to wash up by hand if the energy use is the same? The answer is that the dishwasher uses electricity as its heat source, while most homes use gas for their hot water system. Since each unit of heat produced by electricity generates more than double the CO_2 emissions of the same amount of heat produced from gas (see p.51), then carefully washing by hand will tend to be greener in households with gas boilers. And that's before you consider the energy used in producing, delivering and eventually disposing of the dishwasher – figures that were ignored by the Bonn study.

Cookers

The proportion of your carbon footprint that comes down to cooking depends not just on how much you cook, but also on the type of cooker you use. Gas cookers are far more environmentally friendly than electric ones, for the reasons discussed in chapter four, so if you live on the gas network but have electric hobs, switching over will bring real emissions savings, not to mention a better cooking experience. (Note, though, that unburnt gas leaking from the hobs is mainly composed of methane, which is a greenhouse gas more than twenty times as powerful as CO_2. For that reason, if you have a dodgy ignition button, buy a hob lighter to help limit the amount of gas that leaks out before you manage to light the hobs.)

What about ranges? These larger cookers will tend to take more energy and resources to make than smaller cookers. In use, they're also likely to consume somewhat more energy, given that they often have bigger rings and ovens. However, the difference is unlikely to be large except in the case of cast-iron Aga-style cookers, some of which are designed never to be fully turned off, and kick out heat and emissions all year round – even in the summer. As George Monbiot commented in *The Guardian*, "Compare, for example, the campaign against patio heaters with the campaign against Agas. Patio heaters are a powerful symbol: heating the atmosphere is not a side-effect, it's their purpose. But to match the fuel

Energy-efficiency labels and logos

▶ **EU Energy Label** All retailers must show this label alongside fridges, freezers, washers, tumble dryers, dishwashers, lamps, ovens, light bulbs and air conditioners. Each item is ranked from A (efficient) to G (inefficient) for power consumption under standard operating conditions. Confusingly, though, the scale for fridges and freezers also includes A+ and A++, so don't think you're getting a super-efficient model just because it's marked A. Indeed, these days all washing machines, fridges and freezers tend to be rated C or above so a "middle" rating is actually relatively low. To add to the confusion, the EU is planning to swap the current system for one based on numbers – the result, according to some green commentators, of lobbying by the appliance manufacturers.

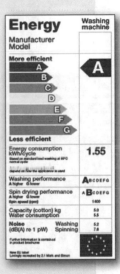

More useful than the ranking is the precise figure provided underneath, usually in terms of kilowatt hours per cycle or per day. If you're trying to balance cost with efficiency, factor in the future cash savings by multiplying the kWh figure by fourteen to give the cost in pence for each cycle or day of use.

The EU label also generally contains other information, such as water usage and performance. Washing machines are given a three-letter rating, in the form AAB, referring to energy efficiency, wash performance and spin performance, respectively.

▶ **Energy Saving Recommended** Administered by the UK's Energy Saving Trust, this logo was developed to point consumers to the most energy-efficient products on the market. It can be found on light bulbs and appliances as well as boilers, heating controls, insulation and more. The criteria are fairly strict. For example, to bear the logo at the time of writing, fridges must be A+ or A++ (25%–45% more efficient than standard A models) and washing machines must be AAA. You can find endorsed products online or by phone:

energysavingtrust.org.uk/Energy-saving-products • 0800 915 7722

▶ **EU Ecolabel** The EU 's Ecolabel flower designates a product that has passed numerous environmental criteria, relating not just to energy use but to expected lifespan, ease of disposability and more. The scheme also covers everything from paints to tissues to computers, but right now few products carry the label in the UK. For more information and a product catalogue, see eco-label.com.

Aga-style range cookers may look homely in a country kitchen, but they're prolific energy consumers

consumption of an Aga, a large domestic patio heater would have to run continuously at maximum output for three months a year." An Aga executive responded to Monbiot's criticism, claiming that an Aga could in fact be a fairly efficient way to cook and heat, but his defence simply didn't stack up.

Whatever you cook on, try to use pans the right size for the hob or ring, and keep lids on when things don't need stirring. According to some estimates, these simple steps alone can reduce the energy required by half.

Microwave ovens

People sometimes assume that microwave ovens are the least green way to cook. In fact, the opposite can be true, depending on what's being cooked and whether the alternative is a gas or electric cooker.

When comparing microwaves to hobs, the microwaves score best for vegetables and other foods that require boiling water if heated up in a pan. In these cases, since a microwave can heat the food directly – without the boiling water – it requires less energy overall. For other foods where everything being warmed up is to be eaten (tinned beans, for example) then microwaves are roughly equivalent to electric hobs, and considerably less green than gas hobs.

What if the alternative is an oven rather than a hob? In this case, the microwave tends to perform fairly well: always better than an electric oven, and for some items, such as potatoes, even better than a gas oven. On the other hand, a microwaved potato doesn't bear comparison with a properly baked one when it comes to taste. Moreover, a really fair comparison of gas ovens and microwaves should factor in the time of year. A large potato may cook with less energy in a microwave than in a conventional oven. But the oven also produces extra warmth in the kitchen – likely to be the room the home's inhabitants are occupying. In the summer this would be a disadvantage, but in the winter it might allow you to turn the central heating off or down, potentially leading to a reduction in emissions as well as a better-tasting potato.

Kettles

In the UK, a nation famously addicted to tea drinking, the kettle is a significant consumer of energy. Electric jug kettles – rare in most of the world – consume more than four billion kilowatt hours of electricity in Britain each year, according to the Market Transformation Programme. That compares to 115 billion kilowatt hours for all domestic electricity use. The impact of electric kettles is even greater than these simple figures suggest, too, as they're often switched on across the country at similar times, creating spikes in electricity demand (see box overleaf).

The most obvious way to reduce the ecological impact of your tea or coffee is to only boil what you need. As kettles have become more power-ful, people have grown less careful about this, as the extra water in the kettle doesn't cause such a long delay. The simple solution is to measure the cold water into the kettle with a cup.

There is, however, another easy way to reduce the carbon footprint of your hot drinks, assuming you have a gas cooker: switch from an electric jug kettle to a stove-top version. The energy used to boil the water will be slightly higher, as some heat escapes around the edge of the kettle, but the emissions and cost will typically be significantly reduced, due to the fact that electricity is so much more polluting and expensive than gas. As a bonus, because a traditional stove-top kettle will whistle when it's ready, you're less likely to forget that you've boiled it and end up reheating the same water half an hour later.

If, on the other hand, you have an electric cooker, then it makes *much* more sense to stick with an electric jug kettle, since the hob will use far more energy to do the same job. Even better, consider trading in your jug kettle for a specialist energy-saving model. These include:

Tefal Quick Cup quickcup.co.uk
The Quick Cup is a dispenser that filters and heats water on demand, so you never need to worry about boiling too much. The device heats each cup's worth of water in just three seconds, and the manufacturer claims energy savings of up to 65%. The

Help shave the peaks

Most of us take it for granted that when we switch on a light or a kettle, the electricity we require will be automatically available. To make this possible, though, the operators of the national grid have to be permanently monitoring electricity demand to ensure that the right amount of power is being generated. At times of highest demand – such as early evenings in the winter – the most inefficient and polluting power stations are rolled out, meaning that each unit of electricity we consume produces extra CO_2.

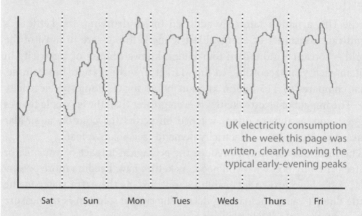

UK electricity consumption the week this page was written, clearly showing the typical early-evening peaks

| Sat | Sun | Mon | Tues | Weds | Thurs | Fri |

In addition, whenever there's likely to be a sudden spike in demand – the classic example being half-time during a cup final, when millions of people simultaneously switch on their kettles and flush their toilets, causing water pumps to kick in – the grid operators have to literally "warm up" fossil-fuel power stations just to make sure they're ready to quickly respond to the boosted demand.

Eventually, new technologies and systems will change the way the grid operates. In a so-called "smart grid", demand will be reduced by various different strategies. For example, "smart meters" will be able to automatically switch off our fridges for a few minutes when there's a spike in demand. Techniques like this will also allow us to increase the proportion of the grid that's powered by renewables, as it will allow grid operators to deal with the intermittent power provided by wind and sun.

In the meantime, there's an environmental case for shifting nonessential power use to times when demand isn't at its highest. Switching on dishwashers and washing machines at night, just before you go to bed, is the most obvious and effective example (as long as you're not a light sleeper with loud appliances). Another example is not turning on an electric kettle the moment a popular television show or sports event ends. Instead, wait five minutes for your hot drink or, even better, switch to a stove-top kettle on a gas cooker.

down side is that part of these savings are achieved by heating the water to only 85ºC, which is good for coffee and hot chocolate, but not ideal for tea.

ECO Kettle www.ecokettle.com

With the ECO Kettle (pictured), you can fill the main water chamber to the brim without feeling bad. When you want a hot drink, simply specify a number of cups – from one to eight – and the kettle will only heat exactly the required volume of water.

Washing machines and tumble dryers

The average UK household uses its washing machine around five times a week, which accounts for around a twelfth of the home's total electricity use. Around 90% of this energy consumption goes towards heating water, so the best way to reduce the impact of each wash – besides using a machine rated A for energy efficiency – is to select the lowest temperature that will get the job done to a satisfactory standard. A 30 degree wash is perfectly sufficient for most non-heavily soiled clothing, and using such a low temperature also has the benefit of being better for your clothes. If there's an economy setting or shorter programme button, be sure to give it go, and always try to run the machine fairly close to full capacity, as it typically takes far more energy to do two half-loads than one full load.

Drying clothes

However you wash your clothes, drying them on a line or rack rather than using a dryer is an obvious way to make substantial energy savings. Dryers typically use considerably more energy than washing machines, with each cycle adding around 30–35p to your electricity bill. That doesn't sound like a huge amount, but used for every wash a dryer can easily add a tenth to your household power consumption.

As climate commentator Chris Goodall points out in his book *How to Live a Low Carbon Life*, basic physics tells us that evaporating water from clothes drying on a rack in the home will cause the air to cool, so in the winter it makes sense to put your rack in an empty room. In the summer, by contrast, putting your clothes rack in the kitchen or draping the wet clothes over radiators and banisters will give you a free air-conditioning effect.

Which detergent?

Traditional detergent manufacturers have for years been criticized both by green groups, for their overuse of environmentally burdensome chemicals, and by animal rights campaigners, for their inclusion of ingredients tested on animals. So should we all switch to "green" brands such as Ecover?

First, let's consider the environmental issue. All things being equal, it clearly makes environmental sense to favour brands that use easily degradable plant-based ingredients rather than less easily degradable and more toxic petrochemicals. Of course, it's also good to support ethically progressive brands. But if you take the view that climate change is *the* key environmental threat, you might decide the most important thing is to save energy by using lower washing temperatures and shorter cycles. And you may find you get better results on those kinds of programmes with a non-eco detergent – especially for clothes that are heavily soiled. If, on the other hand, you're not so fussy about your clothes always seeming cleaner than clean, the ecological detergent will almost certainly suffice.

A more meaningful comparison of different types of washing powder would include a look at the carbon footprint of the production of each brand. Unfortunately, this data isn't available – even for most of the green brands. Ecover is ISO 14001 certified and produced in an environmentally progressive building, but as yet it's impossible to say how its products compare with the mainstream competition in terms of carbon emissions.

Hardline greens point out that, for lightly soiled clothes, it's possible to get by with no washing powder at all, since sweat and much other dirt is water-soluble and removed perfectly well by the warm water and rotating action of a washing machine. This probably explains the success of products with names such as eco-balls, eco-discs or aquaballs that claim to remove the need for detergent by "ionizing" or "magnetizing" the water. Many green consumers swear by them, but sceptics have described them as the washing equivalent of a placebo pill.

As for animal testing, if this is your main priority you should look out for products bearing the Humane Household Products Standard jumping rabbit logo. Even here there's arguably an environmental flip side, however. Current law requires all new cleaning ingredients to be tested on animals before they can be used in products, so to carry the logo a company effectively has to promise never to use any new ingredients – at least not until the laws on animal testing change. That's fine for needless new ingredients designed, for example, to "optically brighten" our whites, but it does mean closing the door on any future chemicals that could help make the products greener.

Lighting

Lighting accounts for almost a fifth of the electricity use in the typical home with non-electric heating. In homes with large numbers of high-wattage bulbs, the figure might be as high as 30%.

There are two obvious ways to reduce the carbon footprint of your lights: turning them off when they're not required, and switching to more energy-efficient bulbs. When you replace an old-fashioned bulb with an equivalently bright compact-fluorescent version, you slash the electricity use by up to 80%. Over the five to ten years that such bulbs typically last, each one can reduce your energy bills by more than £100 (see box) and save a few hundred kilograms of CO_2 emissions.

The cash savings of low-energy bulbs

• An 18W low-energy bulb can last for up to 10,000 hours of use and produces the same amout of light as an old fashioned 100W bulb. In that time, the overall running cost will be around £25 in electricity (10,000 hours x 0.018kW x £0.14 per kWh), plus perhaps £3 for the bulb itself. Total cost = £28.

• A 100W incandescent bulb lasts around 1000 hours. So to produce 10,000 hours of light you'd get through ten bulbs, at around 50p each, and spend a massive £140 on electricity (10,000 hours x 0.1kW x £0.14 per kWh). Total cost = £145.

That's a saving of £117, and a return of 3900% on your £3 investment. The savings per bulb won't be quite so large when the ones being replaced are rated 40W or 60W, of course, or when the low-energy bulb has a shorter life expectancy. Cheaper models often last 6000 hours, rather than 10,000.

The reason compact fluorescent bulbs require so much less energy than their predecessors is that they successfully convert nearly all the power they consume into light, rather than generating heat. By contrast, with an old-style 100W bulb, you're effectively switching on a 20W light source and an 80W electric heater.

Despite their obvious benefits, many people still haven't switched to low-energy bulbs. Over time, everyone will effectively be forced to do so because regular incandescent bulbs are being phased out; 100W globes ceased to be available in many UK stores at the start of 2009, and 60W globes will follow in 2010. This has led to cries of outrage from some quarters, due to a number of largely spurious claims about low-energy bulbs. Let's take a quick look at some of these supposed disadvantages.

Low energy bulbs – myths and truths

▶ **"They produce bad light"** It's true that the colour rendering of low-energy bulbs is slightly different to that produced by incandescent bulbs, but that doesn't mean the light is less good. When US science magazine *Popular Mechanics* asked people to assess the quality of the light from ten bulbs, without telling them which was which, all nine low-energy bulbs tested scored better – for faces and reading as well as general ambience – than the traditional incandescent bulb used as a benchmark.

▶ **"They're not available for most fittings"** In fact, you can get lower-energy replacements for almost all fittings, including halogen spots.

▶ **"They don't work with dimmer switches"** Megaman and other brands produce low-energy dimmable bulbs. They cost more than other bulbs (around £12 each) but will still lead to huge overall cash savings.

Low-energy bulbs are available in a wide variety of shapes – such as this candle bulb

▶ **"They delay and flicker on start-up"** This was once a problem, but decent modern bulbs fire up instantly, without flickering, and reach full brightness within a few seconds (except perhaps in very cold conditions).

▶ **"They're dangerous due to their mercury content"** According to most estimates, there's less mercury inside each low-energy bulb than would be emitted directly into the atmosphere by a coal-fired power station producing the extra electricity required by a traditional bulb. So in fact low-energy bulbs can theoretically *reduce* atmospheric mercury levels. But it's true that you should dispose of them properly. Check your local council's website for advice on where to drop them off.

▶ **"Old-fashioned bulbs let you turn the heating down"** It's true that the heat generated from old-fashioned bulbs isn't always wasted. In the winter, some of that heat may be useful, allowing you to turn the central heating down. However, this only applies during those periods when you have your heating switched on. Moreover, each unit of heat created by an electric bulb will cause more than double the CO_2 emissions of a unit of heat generated by a gas boiler. Furthermore, light bulbs are very often placed high up in a room – on the ceiling, for example – which is a very inefficient spot from which to heat a room, due to the simple fact that hot air rises.

What about halogen bulbs?

Halogen bulbs are a subset of incandescents, and tend to be mid-range performers in the efficiency stakes. Better-quality halogen bulbs are around twice as efficient as typical incandescents. But they're still only half as efficient as compact fluorescents, and any energy savings they do offer are usually offset by the large numbers of bulbs installed in each room. One way to improve things, if you have more bulbs installed than you need, is to remove some of the fittings, or to leave expired bulbs in them. The other approach is to opt for more efficient bulbs. There are three types to consider, all of which are available at GoGreenLights. co.uk:

▶ **More efficient halogens** The most efficient dichroic halogen bulbs offer the familiar halogen brightness, sparkle and dimmability, and they work with any halogen fitting, but they last much longer and use up to 30% less power. Suitable for standard recessed (GU5.3) fittings.

▶ **CFL "halogens"** Megaman and other brands make mini compact fluorescent bulbs designed for use in halogen fittings. These are less sparkly but are available in dimmable versions and have lower energy demands and a longer life than dichroic halogens. These bulbs have a GU10 base for use in non-recessed halogen fittings. (If you want to use them in recessed situations you'll probably need to change your fittings.)

▶ **LED bulbs** Bulbs based on LED – light emitting diode – technology will probably become standard in the future. LEDs use even less power and last even longer than compact fluorescents. Halogen-shaped LED bulbs are already available, but most models currently offer far less light than proper halogens or compact fluorescents, so most brands are best suited for nightlights or mood lighting.

Where to buy low-energy bulbs

Basic low-energy bulbs are widely available in shops and supermarkets, and you may be able to get some for free via your energy supplier (see p.53). For a wide range of energy-saving bulbs, however, you'll need to look online. Any online bulb retailer should offer a decent selection, but you can increase the positive impact of your bulbs by shopping at GoGreenLights. Set up by the author of this book, the site offers a wide range of bulbs, and specifies the cost and CO_2 savings for each. All profits are donated to Oxfam to help support their work on climate change issues in the developing world.

GoGreenLights www.gogreenlights.co.uk

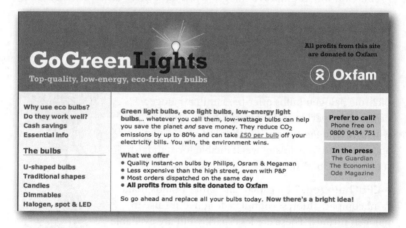

Sunpipes

If you live in a house that needs the lights on during the day, even in the middle of summer, then consider having a Sunpipe installed. These super-reflective tubes carry natural daylight from your rooftop into dimly lit areas, diffusing the light around the room by a translucent ceiling fixture. Sunpipes can provide 100W of light in the winter and up to 500W on a sunny summer day. For more information, see:

Sunpipe www.sunpipe.co.uk

TVs, computers and gadgets

The proliferation of computers, gadgets and super-size televisions is more than offsetting the gains made in the efficiency of white goods and other appliances. Indeed, these consumer electronics already account for around a third of domestic electricity use in British homes heated by gas or oil.

TVs and digiboxes

As a rule of thumb, energy consumption increases with screen size, so the growing popularity of large TVs is directly adding demand to the UK's power grid, much of it during the evening, when the CO_2 emissions per unit are at their highest. Some plasma screen televisions are rated at a massive 500W. Used for four hours each day, such a TV would consume 730kWh a year, which equates to more than 300kg of CO_2 and £100 in electricity bills.

If you're buying a new TV, look for one that carries the Energy Efficiency Recommended logo. For flat-screen models, these will usually be LCDs, which tend to be less power hungry than plasma screens. Note that the logo refers to electricity use per unit of screen size, so a large TV carrying the logo could still consume a fair amount of energy.

Set-top boxes for digital TV reception can also be power hungry, and many rely on inefficient standby systems. So when you do need to buy a new TV, consider one with IDTV (integrated digital television). These save energy by only using one, usually more efficient, power circuit.

Computers and peripherals

Computers and displays vary widely in the amount of energy they consume. One key factor is whether you use a laptop or a desktop. Laptops consume much less energy – partly because they have smaller screens, but also because manufacturers have a strong incentive to optimize the energy efficiency of their laptops in order to give them better battery life. Some laptops consume as little as thirteen watts when idle (turned on but not actually processing any information); some desktops with large monitors use more than twenty times as much. Thankfully, laptops are swiftly becoming more popular than desktops, so this should be at least partly offsetting the increasing number of computers.

If for the sake of budget or any other reason you'd prefer a desktop, consider a super-low-power model from a specialist such as VeryPC. This company produces small, quiet desktop machines that use as little as 24 watts (though that's without a display, of course). They're very affordable, too.

VeryPC www.very-pc.co.uk

As for the Mac versus PC debate, Apple computers score better than average in the energy-efficiency stakes (for details see apple.com/environment). In addition, Apple are one of the few manufacturers to have attempted to understand the cradle-to-grave carbon footprint of their computers. At the time of writing, their basic desktop – the iMac, with a 20" screen – clocks in at just over a tonne of CO_2. Of that, 56% is accounted for by consumer use, 37% by manufacture, 6% by transport and 1% recycling. A MacBook laptop, by contrast, has a carbon footprint of just under half a tonne, half of which is emitted during manufacture, 39% in consumer use, 10% in transport and 1% in recycling.

Whatever computer you use, its energy use will vary in different situations. The more a computer is having to "think" (that is, the more applications it's running, and the more data it's being asked to process) the more power it will require. But in sleep or standby mode the machine will still consume some power, and in many cases (such as with all laptops) the power adapter will use some electricity even when the computer is switched off.

You can make some difference to your computer's energy use by reducing the period of inactivity required to cause the screen, hard drives or whole computer to enter sleep mode. This is done via the Control Panel on a PC or via System Preferences on a Mac. While you're there, consider turning off any screensaver that you may have installed. These no longer have any "screen-saving" role and are purely decorative. Sleep mode typically uses far less power, as it allows the screen to switch off.

Finally, be sure to keep your printer and other peripherals switched off when not in use. Many printers will use up to twenty watts of power even when sitting idle.

Internet access

If you use the Internet, then the energy footprint of your computer may be somewhat higher than just the electricity it consumes directly. Websites, downloads and online applications are typically stored on special computers known as servers. Some servers are located in company offices, while others live in so-called "server farms": buildings containing thousands of computers, each not only consuming electricity but also kicking out heat and therefore demanding energy-hungry air conditioning. Some of the most popular websites – such as Google – have hundreds of thousands of servers around the world.

In 2007, *The Times* caused a stir with an article claiming that carrying out just two Google searches caused the same CO_2 emissions as boiling a kettle. The numbers looked dodgy, however, and Google quickly responded with a much more believable estimate of 0.2 grams of CO_2 per search. If that figure is correct, then even if you searched Google dozens of times a day, Google's servers would account for only a minute fraction of one percent of your carbon footprint.

That said, there's no doubt about the overall significance of the Internet's growing energy needs. According to a study by microchip producer AMD, servers consumed 1% of all the world's electricity in 2005, and the figure is rising fast. Another study, published by McAfee in 2009, concluded that spam email alone causes more electricity use than 2.1 million US homes. For now, though, it's fairly likely that the biggest contributor to the carbon footprint of your Internet access is your wireless router, if you have one. Some of these devices – especially older ones – consume a surprising amount of electricity (a fact reflected in how hot they get), so it makes sense to switch yours off when it's not in use.

Using your computer to fight climate change

Model date and time: 02/09/1922 00:00

Unless there's a reason not to, computers should ideally be turned off when they're not in use for extended periods. But if you can't get into this habit, you could instead dedicate your PC's downtime to climate change research. Supported by the BBC, Climate Prediction is a web-based project that divides complex climate modelling software into tiny pieces manageable by standard PCs (Macs aren't supported at the time of writing). Once installed, your computer will be helping to predict future climate change impacts whenever it enters screensaver mode. For more information, or to get started, see:

Climate Prediction
www.climateprediction.net

Phones, iPods and other gadgets

Portable devices such as mobile phones and MP3 players tend to consume relatively little electricity in use and on standby. Like laptops, they're designed to require as little power as possible in order to help improve their battery life. However, as much as 95% of the total energy consumption of many phones and other handheld devices is accounted for by chargers that are plugged in and consuming power even after the device is recharged or removed. To avoid this waste, it makes sense to follow the standard green advice and unplug your chargers – or switch them off at the wall – when they're not in use. That said, it pays to keep this sometimes exaggerated issue in perspective. As energy expert David MacKay put it in his book *Renewable Energy Without the Hot Air*:

> Obsessively switching off the phone charger is like bailing the *Titanic* with a teaspoon. Do switch it off, but be aware how tiny a gesture it is. All the energy saved in switching off your charger for one day is used up in one second of car-driving. The energy saved in switching off the charger for one year is equal to the energy in a single hot bath.

Devices on standby

Electrical appliances vary widely in how much power they consume when left in standby mode. Some efficient devices use as little as one watt – that's two to three thousand times less than an electric kettle. But others, including some hi-fis, televisions and video game systems, use almost as much in standby as they do in use. Given how many devices are left on standby in the average home, this energy "leakage" adds up. According to the UK Market Transformation Programme, around 5% of the electricity provided to an average home is consumed by appliances left on standby.

Inefficient standby modes are gradually being phased out. In the meantime, getting into the habit of turning appliances off properly can take a small but not insignificant chunk out of your emissions and bills, as the following table makes clear. If you can't get into the habit, help is available in the form of various types of standby busters. The Intelliplug range includes an adapter (pictured) that will automatically switch off computer peripherals when the

PC or Mac isn't in use. Another item in the range will turn off unused TVs, DVD players and hi-fis; it even allows you to switch the various devices back on from cold using their regular remote controls. Ivy Energy, meanwhile, is one of the various companies that produces remote control plugs for turning off appliances around the home at the click of a button.

Ivy Energy www.ivyenergysaving.com
IntelliPlug www.oneclickpower.com

Appliance	Average time in use	Time on standby/ plugged in	kWh used	Percent of electricity used on standby	Annual savings of switching off	Annual CO_2 cut of switching off
TV (80 watts playing, 15 watts on standby)	3 hours per day	21 hours per day	203	57	£13.80	50kg
Video	3 hours per day	21 hours per day	125	74	£11.04	40kg
Battery charger (2 watts)	9 hours per week	159 hours per week	18	97	£2.16	7.5kg
Microwave (8 watts)	20 mins per day	23–24 hours per day	70	98	£8.40	29kg

Get to know your meter

Most people know roughly how much they spend on shopping but have no idea how much electricity they use. That explains why devices which measure or track household power consumption can make such a big difference to behaviour. There are three main types available, all of which have the potential to help you take 10–20% off your electricity bills.

Plug-in electricity monitors

These inexpensive devices allow you to measure the power consumption of individual appliances over time. Start with your fridge or freezer as this is on every hour of every day, so it's easy to extrapolate its yearly consumption from a few days' use. If any of your appliances give figures equal to or greater than the levels in the table overleaf, you might want to consider upgrading to a more energy-efficient model.

Appliance	Electricity use (kWh per year)	Percent of household electricity demand	Cost to run per year	Efficiency savings of a model rated A+	Annual cost savings	CO_2 savings
Lighting	715	27	£86	75%	£64	230kg
Fridge-freezer	650	20	£78	60%	£35	189kg
Dishwasher	410	13	£49	40%	£13	70kg
Washing machine	270	9	£32	30%	£5	27kg

Household power monitors

Household power monitors, available from around £30 to buy or £10 to hire, help you keep an eye on your total electricity consumption. A sensor box clips to the power cable running into your meter. This works out how much electricity you're currently using and beams that information to a wireless display. The display – which you can put in the kitchen, sitting room or anywhere else – shows your power consumption in terms of units, cost and CO_2 emissions. There are many energy monitors of this sort on the market. Popular models include the Owl (pictured), which is available to buy along with various other models at the following website. Before you buy, you could try calling your electricity provider to see whether they offer free or discounted monitors to all customers.

Electricity Monitors www.electricity-monitor.com

Energy trackers

The next step up from a simple energy monitor is a device that will not only display your current electricity consumption, but allow you to track it over time, too. Probably the best model available is the Wattson (pictured). This ergonomic gadget displays your electricity use in terms of how much you'd be spending in a whole year if you left everything running at its current level. For the really committed, the Wattson can also be connected to a computer, allowing the associated software (called Holmes, of course) to track and even graph your power use over time. Wattsons are available for around £100 from:

DIYKyoto www.diykyoto.com

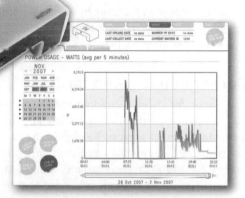

Basic household energy monitors such as the Owl (left) provide an inexpensive way to keep track of your electricity use in real time. More up-market models, such as the Wattson (below left), can also be set up to track your usage over the days or weeks using special software (below).

Smart meters

In the long run, devices such as Owls and Wattsons will probably be rendered redundant by smart meters in every home. These will not only feature a power monitor but will be able to transmit power consumption to the utility company in real time. This will end the need for meter readings as well as allowing for an electricity price that changes minute by minute to incentivize people to switch off at times of peak demand. Best of all, smart meters have the potential to change the behaviour of household devices according to power availability. For example, fridges might be automatically switched off for a few minutes during a spike in demand to avoid the need for an extra power station coming online. Some countries, such as Italy, already have smart meters in many homes. The UK government plans to install them in all homes by 2020.

Green electricity & gas tariffs
Is it worth signing up?

A green electricity tariff is a power package that claims to offer some kind of environmental benefit. There are many such tariffs available in the UK, both from major utilities and from green specialists. Some are simple and not very ambitious: the supplier promises a token gesture such as planting a tree on behalf of each new customer. However, other tariffs offer – either explicitly or implicitly – to provide you with electricity from renewable sources. These schemes sound very tempting but, as we'll see, the advertising is usually somewhat misleading.

In addition to regular green tariffs, there's also the option of a tariff geared towards cooperative values and social justice, or a tariff that encourages you to save energy by giving discounts when you reduce your energy consumption. These alternative schemes – discussed on p.93 – are arguably more attractive than conventional green tariffs.

Conventional green tariffs

The market for green electricity tariffs has for years been highly confusing, with some tariffs offering far less environmental benefit than the marketing material suggests. Friends of the Earth gave up ranking green tariffs in 2004 due to the complexity of working out which were and weren't providing any real carbon savings. And a 2006 report by the National Consumer Council concluded that none of them could really claim to be offering substantial environmental benefits.

One complication comes from the fact that UK law requires all electricity providers to buy a proportion of their power from renewable sources. This rule, known as the Renewables Obligation, requires an ever-rising proportion of UK electricity to be derived from renewable sources. At the time of writing, the Renewables Obligation is 9.1%, rising about 1% per year. The cost of meeting – or trying to meet – these targets is funded by a surplus on all electricity tariffs, green and regular alike.

This grid-wide requirement is a good thing, but it leaves open the possibility that companies can sell green tariff customers the renewable energy that they were going to have to have in their portfolio anyway due to the Renewables Obligation.

Even if your supplier buys more than the legal minimum from renewable sources, it doesn't necessarily make them more green. For each unit of renewable energy that a company generates it gets a Renewables Obligation Certificate (ROC); if it generates more than the legal requirement, it can sell the extra ROCs to other companies, which then avoid generating the green power themselves. For this reason, the National Consumer Council and some other groups have argued that the only green tariffs that make a difference to the amount of renewable power being generated are those from companies that buy more than the required percentage of green electricity and then destroy ("retire") some of the extra ROCs to help drive up their price and incentivize green generation. Some specialist power providers, such as Good Energy, have responded to these concerns by retiring a few percent of their ROCs. Today, however, it's not even clear that this makes any difference, because the barriers to new renewable energy capacity are not lack of demand but planning and technical obstacles. Retiring ROCs can't help much with those kinds of barriers.

Due to these kinds of complications, Ofgem finally decided to create new guidelines for green tariffs in 2009. These have helped clarify things a little, but not completely. To get the Ofgem seal of approval, the environmental benefits of the scheme have to be completely unrelated to the electricity being purchased by the customer. Instead, the power company has to either buy carbon offsets (at least one tonne of CO_2 savings per customer) or contribute towards community renewable or energy-efficiency schemes, such as putting solar panels on school roofs.

On the one hand, this seems a crazy system. You might as well just stick with a regular tariff and pay £8 to a carbon offset service to soak up a tonne of CO_2 in the developing world. On the other hand, at least the rules are now clearer, and in one sense switching to overseas offsetting makes perfect sense given that the UK's power sector has capped carbon emissions anyway (see p.34).

At the time of writing, all the big six energy providers – British Gas, EDF, E.ON, NPower, Scottish Power and Scottish & Southern Energy – have signed up for the guidelines for their green tariffs, along with eco specialist Good Energy. But Ecotricity, another specialist, has criticized the guidelines and decided to continue on the same path they were on before – trading ROC certificates and using the money to help support their wind turbine business.

With such a complex set of factors to consider, it's not easy to know the real impact of the various green tariffs. Moreover, there's the wider behaviour of the suppliers to consider. A tariff from a green specialist might offer no greater environmental benefits than one from a major power supplier such as E.ON. But whereas the green specialist is likely to be actively doing everything it can to expand renewables – including small-scale ones that might bypass planning delays – E.ON is trying to get permission from the government to build a large coal-fired power station in Kent.

This, along with the fact that you'll be registering your concern about climate change, is the only real reason to sign up for a green electricity tariff. The important thing, if you do sign up, is to not succumb to the fairly common misconception that your electricity supply will somehow be greener and therefore it doesn't matter how much you use. That simply doesn't add up.

There are around twenty green tariffs on the market. Here are four of the ones which have decent grounds for claiming to be among the best.

British Gas Zero Carbon www.britishgas.co.uk • 0845 456 9550

British Gas took the market by surprise when it announced its zero-carbon electricity tariff, which offers 12% ROC retiral, a contribution to a green fund and a carbon offset package equal to all your gas and electricity use (the scheme is only available if you sign up for both gas and electricity). Annual electricity price for a typical household: £468.

Ecotricity www.ecotricity.co.uk • 0800 032 6100

Ecotricity focuses primarily on developing new wind farms. They do not retire any ROCs, and aren't signed up to the Ofgem guidelines, but they do have a good track record of actually getting turbines up and working. Annual price for a typical household: £468–488 depending on the package.

Good Energy www.good-energy.co.uk • 0845 4561 640

Has a portfolio of 100% renewable electricity and retires 5% of its ROCs. It also buys energy from small generators through its Home Generation scheme, and has signed up for the Ofgem green tariff guidelines, which means it will have to also contribute to community schemes or offset projects. Annual price for typical household: £540.

Green Energy UK www.greenenergy.uk.com • 0845 456 9550

Invests in small-scale renewables as well as "low impact" combined heat and power schemes (CHP). Annual price for typical household: £520–560, depending on the package.

Socially equitable gas and electricity

Ebico is a nonprofit organization that focuses on social equity as well as the environment. With conventional electricity and gas suppliers, the poorest members of society end up paying the most per unit. This is because low-income customers often use expensive pre-pay meters, lack bank accounts enabling them to benefit from direct debit savings, or pay a disproportionate amount via standing charges. Ebico's Equipower and Equigas schemes offer one price per unit, regardless of the payment method. Yet, because it's non-profit-making, the company is able to offer extremely competitive rates and good customer service.

Best of all, for green-minded customers Ebico offers a meaningful environment option ("Equiclimate") whereby it purchases EU Emissions Trading Scheme carbon credits to match your total electricity and gas use and withdraws them from the market, hence forcing down the overall emissions of the power sector.

Ebico www.ebico.co.uk • 0800 458 7689

Energy saving tariffs

Southern Electric's Better Plan tariff takes a completely different approach to environmentally friendly energy supply. The basic premise is that the greener you become, the less you'll pay for your electricity and gas. Customers qualify for a discount of £15 when they reduce energy bills by 10% in a year. Further discounts are offered for choosing paperless billing and for improving your home's energy efficiency: £10 for buying A-rated appliances and £20 for upgrading an old boiler or insulating your roof or walls. An energy monitor is also included with the package.

If by signing up for this tariff you succeed in motivating yourself to take all the recommended steps, the impact on your carbon footprint could be substantial – far larger than with a conventional green tariff.

Southern Electric www.southern-electric.co.uk

More information and switching

For more information on the companies mentioned above and other power providers, including their respective fuel mixes, rates of ROC retiral and prices, visit Green Electricity Marketplace. There are other comparison services out there, too, but they haven't all grasped the complex ins and outs of the green tariff market.

Green Electricity Marketplace www.greenelectricity.org

If you do decide to change energy supplier – or indeed phone company, broadband or any other service – you might want to increase the positive  impact by doing so via Switch and Give. Websites such as this, which compare household services and enable you to switch, typically receive a commission from whichever supplier you sign up with. Unlike other such services, however, Switch and Give donates £10 of this commission to a charity of your choice.

Switch and Give www.switchandgive.com

Installing renewables at home

Clean heat and microgeneration

With high energy prices and an ever-greater sense of urgency surrounding climate change, there's never been a better time to investigate the various renewable energy systems that can be installed at the household level. For heat, the main options are solar water heating, heat pumps that exploit warmth from the ground or air, and boilers that burn wood or biomass pellets. For electricity, there are solar photovoltaic panels, micro wind turbines and, for those few with a suitable river flowing through their land, micro-hydro. A final option is micro combined heat and power – not a renewable system as such, but a potentially carbon-saving home-generation technology nonetheless.

This brief chapter takes a look at the pros and cons of each of these systems. As we'll see, the upfront cost is usually fairly high, so it makes economic as well as environmental sense to focus first on insulation and energy reduction – as described in the previous chapters. Nonetheless, if you have money to spend, a home renewable system can offer a good return on your investment. You might find that you'd save more in reduced energy bills than you'd receive in interest if you just left the money in a savings account, especially if you pay a high rate of tax on your savings. Moreover, the initial outlay may be partly or entirely offset by the value added to your home, especially if you can reduce the cost with an installation grant (see p.52).

Energy payback

One concern occasionally voiced about renewable energy systems is that the fossil fuel energy used during their manufacturing might outweigh the clean energy they create when in use. In truth, this shouldn't be much of a concern: all the devices discussed in this chapter are likely to displace far more dirty power than was used in their production. It's true that, some time ago, certain solar photovoltaic panels took more than a decade to repay their manufacturing energy, but today the figure is down to just a few years, compared to a lifespan of at least a quarter-century.

Renewable heat and hot water
Solar hot water, heat pumps and biomass

Renewable heating systems can be attractive for many homes but especially those which are off the gas network. That's because, as we saw in chapter five, heating your home and water with oil or electricity is much more expensive and polluting than using gas.

Solar water heating

Though most people associate solar power with generating electricity, it's also possible to convert sunlight directly into heat. The typical setup is a roof-mounted solar collector panel that channels the sun's energy into your hot water tank. (As discussed on p.65, it's also possible to use a solar system to warm the cold water feed of a combination boiler.) Solar water heaters have been around for decades and there are around 100,000 installations in the UK.

There are various types of solar collector available, with efficiency increasing with price. Beware of junk mail and door-to-door salesmen that promise savings that are too good to be true, and always get several quotes before committing to an installer.

▶ **How much heat?** In the UK, a typical system will provide all your hot water in the summer and about a third of your total annual demand. You'll typically do better than that in the sunny Southwest and less well in cloudy northern Scotland.

▶ **Space requirements** A typical three-bedroom house would need three or four square metres of roofspace facing southwest, southeast or, ideally, south. You'll probably also need space for a hot-water cylinder, if you don't have one already.

Solar Twin

A Solar Twin hot water collector in Fife

▶ Cost and grants

Expect to pay around £3000–5000, depending on the type and size of collector. (Installation costs are reduced if combined with other roof work.) The Low Carbon Buildings Programme offers a £400 grant towards the cost.

It is possible to install your own panels, which could reduce the outlay by £1000–2000. However, a DIY installation is not eligible for a grant, and you'll pay full VAT for the panels rather than the reduced 5% level paid by accredited installers.

▶ **Payback period** Savings are best in areas off the gas network. You could expect a payback in 25 years when compared to electric water heating. The payback could be much quicker with a DIY installation and/or in a home with high hot-water demands. If you're on the gas network, on the other hand, the payback might take much longer.

▶ **Planning permission** Not required except if you're in a conservation area and the collector will be visible from the road.

▶ **Maintenance** Virtually none is required. You may need to replace the antifreeze after about five years. Other than that, just check annually for debris on the collectors, cleaning them when necessary with soapy water.

▶ **Worth looking into?** Perhaps. It makes most sense for those with electric water heaters and especially those with high levels of hot-water demand in summer (such as B&Bs and campsites). For more information, see:

Solar Trade Association www.greenenergy.org.uk/sta

Heat pumps

Ground-source heat pumps

Ground-source heat pumps extract warmth (ultimately solar energy) stored in the ground. A length of plastic pipe is buried in your garden or land and filled with a mixture of water and antifreeze. This liquid absorbs heat from the ground – which never goes below a few degrees Celsius even when the air is freezing – and an electric compressor raises the temperature to a useful level. The heat is then distributed around the home, typically via underfloor heating since the system is much more efficient at producing lots of low-level heat than it is at generating a smaller amount of the higher-level heat typically used by radiators.

The pump and compressor consume some electricity to operate but as long as the home is well insulated then the total energy cost and emissions will be far lower than with a typical heating system – especially a heating system based on electricity, oil or solid fuel. (Of course, if you combine a ground-source heat pump with solar panels or a wind turbine, then you can create a completely renewable system.) In badly insulated homes, the benefits are greatly reduced, because the pump has to work much harder, consuming valuable electricity.

thehouseofyeager.com

The coil for a ground-source heatpump being laid in a garden in the US

If you use air conditioning in the summer, you might want to investigate reverse-cycle heat pumps that can provide both heating and cooling. Unfortunately the UK grant scheme does not cover these.

▶ **How much heat?** You could produce all your space heating with a ground-source pump, and much of your hot water.

▶ **Space requirements** The ground loop needs a trench 75–100m long and 1–2m deep for a typical house. Vertical systems based on deep bore holes are possible and take up less space, but they're also more expensive. The pump itself is a fridge-sized box (see picture).

▶ **Costs and grants** A typical 8kW system costs £6000–12,000, not including the underfloor heating. A total outlay of £10,000–13,000 would be realistic for an average house. The Low Carbon Buildings Programme offers a grant of up to £1200.

▶ **Payback period** Savings are best in areas without mains gas. If you replace an electric heating system you could reduce your bills by up to two thirds, enabling the system to pay for itself in ten to fifteen years.

▶ **Planning permission** Not required.

▶ **Maintenance** The pump should be serviced from time to time.

▶ **Worth looking into?** For homes off the gas network, then perhaps – especially those being built from scratch or undergoing total renovation. This way a trench can be excavated during the building works and underfloor heating easily incorporated. For more information, see:

Ground Source Heat Pump Club www.nef.org.uk/gshp/gshp.htm

Air-source heat pumps

Air-source heat pumps work in a similar way to ground-source models though they exploit the low-level heat in the air rather than the ground. They're far less efficient than ground-source pumps for cooling (since the air gets warm on hot days), but they're almost comparable for heating and hot water. Best of all, they're slightly cheaper to buy (£6000–8000) and don't require the space, hassle and expense of digging a trench. This makes them much more suitable for flats and urban houses. One down side is that some air-source pumps can be slightly noisy.

A box around the size of a storage heater is placed on the exterior of the home. The heat is either piped to the heating system (ideally underfloor) or distributed directly into the home via fans. As with ground-source pumps, air-source heating is most suitable for very well-insulated homes. The maximum grant available is £900, and by the time you read this planning permission probably won't be required except in conservation areas. Manufacturers include:

Heat King www.heatking.co.uk

Biomass heating and hot water

Biomass includes any biological material. In the context of heating, it usually refers to wood – in the form of logs, chips and pellets (compressed sawdust) – though cereal grain can also be used as a fuel. When wood is burnt, it does produce CO_2 but this is absorbed by new trees planted to replace the mature ones felled. In theory, this makes wood a carbon neutral fuel. In reality, some additional CO_2 emissions result from the harvesting and transporting of the wood, but the overall carbon footprint is still typically *far* smaller than with a fossil-fuel heating system.

Open fires and log-burning stoves

Traditional open fires are the least efficient way to create heat from wood. Not only do they give out relatively little heat per kilogram of fuel; they also produce large amounts of soot, which is itself a major contributor to the greenhouse effect. For both these reasons, and to avoid having a draughty exposed chimney when the fire isn't in use, a wood-burning stove is a far greener choice. The best stoves provide a room with many times more heat than an open fire does from the same amount of fuel, and some burn so cleanly that they can be legally used in smoke-free zones. Any dealer will be able to tell you which of the models they sell are smokeless-approved, but for a full list see airquality.co.uk/archive/smoke_control.

Many large stoves come with an optional back boiler that enables the device to provide hot water and central heating. If you're opting for one of these, then you could qualify for a Low Carbon Buildings Programme grant of up to £1500.

Log-burners cost around £500–1500, depending on size and quality and on whether a back boiler is included. You'll also need to budget around £500 for installation, and in some homes the same again for a new chimney lining.

Automated pellet and wood-chip burners

Automatic heating systems that burn wood pellets or chips require far less attention than a log-burning stove. You fill a hopper with fuel and it will take care of itself for a few days, responding to your heating controls and thermostats. For this reason they're much more viable as a primary heating source than log stoves, which are typically used alongside a fossil-fuel system. Pellet and wood-chip burners are yet to catch on in a huge way in the UK, but they're a proven technology: Austria alone already has more than 100,000 installations.

▶ **How much heat?** A small pellet-burning room heater can be used to provide warmth to a single room, while a larger pellet boiler system can easily produce enough heat and hot water for a standard

Some pellet burners are designed to go in a cupboard or utility room, but many others, such as this one from Artel, can make a focal point in the living room

home. Wood-chip boilers are more powerful still and best suited to large buildings with plenty of space, such as farms.

▶ **Space requirements** Wood has a lower energy content than fossil fuels, so you need more of it to provide the same amount of heat. A room heater fuelled by pellets might only be required for a few hours a day during the winter months. This would add up to an annual heating demand of around 1800kWh, which would require 360kg of pellets. So a year's supply could be accommodated by a typical garage. (The room heater itself will only be around the size of a typical wood-burning stove.)

For larger systems providing hot water and central heating, space becomes a more pressing issue. A large old house with a total water and heating demand of 25,000kWh would require a tank roughly three cubic metres in size if powered by oil, compared to ten for wood-pellet storage and thirty-eight for wood chip. You'll also need space to house the boiler and hopper (generally about double the size of an oil boiler) and a flue.

▶ **Costs and grants** Stand-alone room heaters cost around £1500–3000, while pellet boilers big enough for a typical house go for £5000–10,000. A 100kW wood-chip boiler for a farmhouse and offices might cost £30,000. Unlike other renewable technologies you will need to budget for fuel. However, pellets and chip are fairly inexpensive compared to other fuels (see chart).

The Low Carbon Buildings Programme offers a grant for room heaters and systems with automated wood pellet feed of up to £600, limited to 20% of the total cost. For wood-fuelled boiler systems (regardless of size) the maximum grant is £1500, limited to 30% of the total cost.

▶ **Maintenance** You'll have to empty the ash pan from time to time. This might be weekly for large boilers, or just a few times

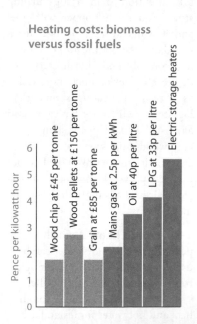

Heating costs: biomass versus fossil fuels

a year for room heaters. The burner should be cleaned once a year; this can be taken care of as part of a maintenance contract with a fuel supplier or boiler manufacturer.

▶ **Payback period** A large farmhouse with outbuildings and an annual heating requirement of 75,000kWh might pay back the cost of conversion from oil to wood-chip in just a few years. But in a typical home, a pellet boiler would take at least fifteen years to pay for itself.

▶ **Planning permission** Not required unless the flue exceeds the roof height by more than one metre – or unless the flue is installed on the principal elevation of a house in a conservation area and is visible from a road. Note, though, that few pellet or chip burners currently on the market are eligible for use in smoke-control zones.

▶ **Worth looking into?** If you live in a rural area off the gas grid, have ample space to accommodate the boiler and fuel and don't mind a hands-on heating system, then yes. It could slash your carbon footprint and, over the system's lifetime, save you some money, too. There are many suppliers of wood-based heating systems in the UK, including:

Rural Energy Trust ruralenergy.co.uk • 01664 454 989
Wood Energy woodenergyltd.co.uk • 01398 351 349
Econergy woodenergy.co.uk • 0870 0545 554
Talbotts talbotts.co.uk • 01785 213 366

You can find your nearest wood fuel supplier at:

Log Pile www.logpile.co.uk

Wood-fired district heating

Just as it's possible to heat individual homes with wood, it's also possible to heat entire villages or towns. In some areas of Scandinavia most houses don't heat their own water and radiators. Instead, hot water is piped in from a small electricity power plant nearby. This is much more efficient than the traditional set-up, whereby power stations throw away massive amounts of heat via cooling towers while local houses burn gas to keep warm. Even better, these small electricity plants are well suited to being powered partly or entirely by wood, meaning that the overall carbon footprint is tiny. The multiple benefits of this wood-fired "combined heat and power" approach are gradually being noted by policymakers around the world, and some climate experts have called for a massive roll-out of the system in the UK and the rest of Europe.

Renewable electricity systems
Micro renewables: solar PV, wind, hydro

If the idea of turning your home into a miniature power station appeals, then explore solar photovoltaics or – if you have a suitable plot of land – the micro versions of wind and hydro generators. It is possible to become truly self-sufficient in electricity with an off-grid system involving a large battery setup. But this is far more expensive and far less environmentally friendly than hooking up your micro-generator to the grid. This way, when your panels or turbines are producing more power than you need – such as when you're out at work – then the excess power is automatically fed into the grid, reducing the need for fossil-fuel generation elsewhere. This works well, though it means that the financial benefits of owning the system can be closely bound up with the price you receive for each unit you feed into the grid.

In addition to selling electricity to the grid, home generators can also earn Renewable Obligation Certificates (see p.91), which can be auctioned at www.e-roc.co.uk for as much as £50 each, depending on the market at the time. Arguably, though, selling the ROCs reduces the environmental benefit of installing the system, as the company buying the certificates will be able to produce less clean energy itself.

Buy-back deals and feed-in tariffs

At the time of writing, probably the best deal available for micro generators is Good Energy's HomeGen scheme. This tariff simply pays you 15p for each kilowatt hour of electricity generated from solar panels or wind turbines, regardless of whether you use it yourself or feed it back into the grid. As part of the package, Good Energy take any ROCs you earn, but that at least saves you selling them. It's a fairly good deal, especially if you use a lot of the power you generate yourself.

All of this may change in 2010, when the government has promised to roll out a national feed-in tariff. The details are yet to be finalized, but the policy should guarantee a higher price for the power generated by home renewables and fed into the grid. The rate in Germany, the nation which spearheaded the feed-in-tariff approach, is a massive 57.4 Euro cents per kWh. If the UK offers anything close to that amount the financial benefits of installing solar and other micro renewables will shoot right up.

It's likely that, for home generation, the feed-in tariff will replace ROCs as well as the grants available via the Low Carbon Buildings Programme.

Solar photovoltaics

Photovoltaic (or PV) systems involve two or more thin layers of semiconducting material – usually silicon – which generate electrical charges when exposed to sunlight. The voltage from a single cell is low, so many cells are connected to form a solar panel or module. These panels are in turn combined into a solar roof or some other kind of "array". The UK still has a relatively small number of arrays but Germany, where there has been a feed-in tariff for some years, has hundreds of thousands.

Solar roof tiles by Solarcentury, seen towards the top of this roof in Allington, Lincolnshire are designed to blend in with existing materials. See solarcentury.com for more information.

▶ **How much power?** A system rated at one kilowatt might produce 800 kilowatt hours of electricity per annum, depending on your location in the country – the intensity of solar energy is much greater in Dorset than in Dumfries. That would be sufficient to provide the baseload of your electricity use, or about 25% of your total yearly requirements. In most cases, the limitation for bigger systems is roof space.

▶ **Space requirements** A 1kW system requires an area of ten square metres of sloping, south-facing roof. It is possible to install panels on other aspects and angles though the amount of power produced will be reduced (around 12% lower if horizontal, 18% lower if on a sloping roof facing east or west, and 29% lower if on a vertical south-facing wall).

▶ **Costs and grants** Expect to pay around £8000–10,000 for a 1kW system. The Low Carbon Buildings Programme offers a maximum grant of £2000 per kW installed, up to a maximum of £2500 and limited to

30% of total installation costs. So if you get a grant your outlay may be as low as £6000.

▶ **Payback period?** A 1kW system generating 800kWh per year might yield an annual electricity saving of around £175, plus £120 via a buy-back tariff. If you get a grant and pay £6000 up front, then the return on investment would be around five percent, and the payback period would be twenty years – only five years less than the warranty period for the panels. In reality, however, the expected feed-in tariff will almost certainly increase the financial benefits and the panels will typically outlive their warranty period by some way. (The first commercial panels ever installed, in Tokyo in the 1960s, are said to be still producing plenty of power after more than forty years.) It's worth considering the substantial value that will be added to your home, too.

▶ **Maintenance** Virtually none is required, though it's a good idea to check annually to see if any debris has fallen on the PV cells. The panels can be cleaned with soapy water and a soft brush.

▶ **Planning permission** Not required unless you're in a conservation area and the panels will be visible from the road.

▶ **Worth looking into?** Perhaps, especially if you have money in the bank, an unobstructed south-facing roof in a sunny part of the country, and no plans to move home. For more information on solar photovoltaic systems, see:

British Photovoltaic Association www.greenenergy.org.uk/pvuk2

Micro wind turbines

Wind turbines create electricity from the kinetic energy of moving air. The power output of each turbine depends principally on three factors: the length of the blades, the wind speed (which varies from area to area as well as by height) and whether there's any air-flow obstruction from other buildings or trees. In short, the bigger and higher up the turbine, the better the result in terms of power generated per pound invested. Because of this, the ideal way to get involved with the installation of wind turbines is not to put one up yourself, but to join a community-owned scheme erecting commercial-scale turbines (see p.108).

Unfortunately, community wind hasn't yet taken off in the UK in the same way that it has in Denmark and elsewhere. The next best option, if you happen to own a suitable piece of land, is to install a stand-alone turbine mounted on a mast. These range in capacity from around 600 watts to more than 20 kilowatts, with the masts ranging in height from around 3m to 15m. One of these could generate more than enough to match your total electricity consumption.

If, like most people, you don't have a suitable patch of windy ground to mount a turbine on a mast, then you could consider a truly micro roof-mounted model. The problem with these is that they don't tend to produce very much power – especially on low buildings and in built-up areas where there's likely to be a lot of turbulence. The technology is gradually improving, however, and in a windy spot with a better-than-average turbine you might generate a decent chunk of your power requirements.

▶ **How much power?** As a guide, a small 1kW roof-mounted turbine with a 1.75m blade might realistically produce about 650kWh per year at an average wind speed of 4.5 metres per second. That's about 20% of a typical household's needs. Stand-alone turbines are far more powerful, though their outputs range widely according to position and height. An appropriately sited 6kW turbine with a blade diameter of 5.5m, raised

Two Swift Turbines on a domestic building in Berwickshire

15m above the ground, should be capable of producing 7500kWh from an average wind speed of 5mps. This is about double the needs of a typical home.

▶ **Space requirements** You don't need loads of space, but best results are achieved at 10m or more above surrounding buildings and trees – which is usually impossible with a roof-mounted turbine. In mounting any turbine, avoid sites with excessive turbulence, which will reduce performance and shorten the device's working life.

▶ **Costs and grants** You should expect to pay about £3000 per kW, including the turbine, mast, inverters and batteries (if required). Large, stand-alone turbines may also require foundation work, which can cost an additional £3000. The Low Carbon Buildings Programme offers a grant of £1000 per kW installed, up to a maximum of £2500 and limited to 30% of the total project costs.

▶ **Payback period** You might break even in fifteen to twenty years with a stand-alone turbine. Most smaller units struggle to pay for themselves during their lifetimes, though a steep rise in energy prices – or a generous feed-in tariff in 2010 – could change this.

Community wind

In Denmark, more than a fifth of wind farms are owned by local residents and although the UK is a long way behind, quite a few wind co-ops have appeared, with many more in the pipeline. Such schemes typically have a minimum and maximum investment, with first refusal offered to people living in the vicinity of the project. This way, the benefits can be realized by as many people as possible, each member has an equal say, and the local community develops a real sense of ownership over the wind farm.

The Baywind community wind farm in Cumbria (www.baywind.co.uk) is the oldest example in the UK. It was set up in 1997 and has 1300 members, 43% of whom live in Cumbria and Lancashire. The cooperative raised nearly £2 million for six turbines on two sites, which produce enough electricity to power 1800 homes. Baywind has frequently provided investors with annual returns of 7% gross and also promises to return the initial capital sum after 25 years. An additional benefit is the tax relief that members get in the form of the government's Enterprise Investment Scheme.

For more information about community wind projects, see:

Energy4All www.energy4all.co.uk

▶ **Maintenance** The service schedule will be specified by the manufacturer. At the very least an annual inspection should be performed to check for physical damage and wear.

▶ **Planning permission** By the time you read this, it's likely that there'll be no requirement to seek planning permission for either roof-mounted or free-standing turbines, except in conservation areas. At the time of writing, however, it's still necessary to consult your local authority.

▶ **Worth looking into?** Probably not for rooftop installations, though perhaps for stand-alone turbines, depending on your budget and the average windspeed around your property. You can find windspeed estimates for your postcode here:

UK Wind Speed Database www.bwea.com/noabl

For more information on micro wind turbines, see:

British Wind Energy Association www.bwea.com/small

And for some examples of UK-made wind turbine models, visit the following manufacturers' websites:

Windsave www.windsave.com
Proven www.provenenergy.com
Swift www.renewabledevices.com/swift

Micro hydro

If you happen to have a river or stream on your property, or you live near to one, you could investigate micro hydro. In such systems, water flowing steeply downhill – over a natural waterfall or man-made weir – is diverted via a pipe called a penstock. This directs the water through an enclosed turbine, which rotates to produce electricity. After leaving the turbine, the water is discharged back into the river.

▶ **How much power?** Micro hydro systems can provide a great deal of power, depending on the height that the water falls (the head) and the volume of water passing through the turbine each second (the flow rate, measured in cubic metres per second). The flow of the water will vary throughout the year so a 10kW system might be able to achieve its full potential only 40–50% of the time. This would still enable the production of 35,000kWh per year, enough electricity for ten homes.

▶ **Costs and grants** Low-head schemes (with a height of 5–20m) usually cost about £3000–4000 per kW. So a 10kW system would come in at around £30,000–40,000. The Low Carbon Buildings Programme offers grants of £1000 per kW installed, up to a maximum of £2500 or 30% of installation costs.

▶ **Payback period** This depends on the price you get for any power you export to the grid. At 8p per unit, a 10kW system producing 35,000kWh per year would bring in an annual income of £2800, plus another £1000 for ROCs. A £40,000 project would therefore pay back in around six to ten years, depending on how much of the power you consume directly and how much you export to the grid. This payback time could be reduced further if a favourable feed-in is implemented in 2010.

▶ **Planning permission** Micro hydro schemes need an abstraction licence from the Environment Agency.

▶ **Worth looking into?** Perhaps – particularly if your land has an old water mill on it, as this will significantly reduce infrastructure costs. There are 30,000–40,000 disused water mills in the UK, only a handful of which are currently used for power production. On a smaller scale, hilly areas with spring-fed streams and a head of just 1m can also be suitable. For more information, visit:

British Hydro Power Association www.british-hydro.org

Micro combined heat and power
Fuel-cells & Sterling-engine CHP

A final way to create heat and electricity in a home is with a micro combined heat and power unit. These are yet to be rolled out widely in the UK, though they're likely to become more common over the next few years. In most cases, micro-CHP units are designed to run on mains gas, so this isn't a renewable technology, but it does have the potential to reduce energy bills and emissions.

The basic idea of micro-CHP is that it's more efficient to produce electricity and heat together, in situ, than it is to get electricity from the grid and to create heat separately. There are two reasons why this is the case. First, electricity generation always produces heat as a byproduct. At a conventional power station, this heat is treated as a waste product and expelled via a cooling tower. A micro-CHP unit, by contrast, can use it for heating water and radiators. The second benefit is that, since the electricity is produced at the point of demand, there are fewer transmission losses, which reduces energy waste and CO_2 emissions even further.

The precise benefits depend on the pattern of your heat and electricity consumption, since micro-CHP systems are at their most efficient when producing heat and electricity continuously. As with all micro-generation technologies, the financial benefits also depend on the price received for any electricity exported back to the grid.

There are two main types of micro-CHP system:

▶ **Sterling engine micro-CHP** In this system the gas is burned, as in a regular boiler, but most of the heat created is converted to electricity thanks to an integrated Sterling engine. A typical model might produce around 2400kWh of electricity each year, as well as 18,000kWh of heat. Two brands have undergone testing in the UK: WhisperGen by Powergen and Microgen by British Gas. If finally released commercially, these devices are expected to be available to residential customers for around £500–1000 more than a normal boiler, with a payback time of around three to four years.

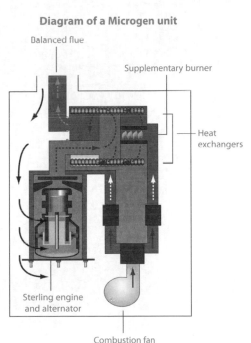

Diagram of a Microgen unit

Balanced flue

Supplementary burner

Heat exchangers

Sterling engine and alternator

Combustion fan

▶ **Fuel-cell micro-CHP** Fuel cells are essentially batteries with a replenishable fuel source. At present, they're most commonly associated with futuristic vehicles but they arguably have more potential for home energy. A typical system will convert the methane in mains gas into a mix of hydrogen and carbon monoxide. The electrons from the hydrogen are used to create a flow of electricity, with the only byproducts being heat and water. Despite successful trials in Japan and elsewhere, domestic fuel cells are yet to catch on in the UK. A planned trial roll-out funded by E.ON could change that.

Waste & recycling

How to reduce, reuse and recycle

British households produce a staggering 25 million tonnes of refuse every year. That's around half a tonne per person. Add in our share of the country's commercial, industrial and agricultural waste and the figure rises to 4.3 tonnes per head – approximately sixty times a typical person's body weight. Of our domestic waste, recycling and composting rates have shot up from around 8% in 1998 to around a third in 2008. That's a huge improvement, though as we'll see Britain is still a long way behind some best-performing countries.

Various EU laws and UK targets will force our recycling figures to continue climbing steeply over the coming years, but in the meantime there's much we can do on an individual level to reduce the amount of rubbish that we generate. In doing so, we can help cut back on the greenhouse-gas emissions that result from the production of packaging and other items we regularly dispose of, from the decomposition of biodegradable waste, and from the collection and processing of our refuse.

In trying to reduce the amount we throw out, the old green mantra of the three Rs – reduce, reuse, recycle – holds as strong today as ever. This chapter takes a look at each "R" in turn, after quickly examining what we throw out and what happens to the waste that we don't recycle.

What we throw out...

The data on the materials we recycle is fairly good, but there's less information available about the total makeup of our waste, including both recycling and garbage. The following chart shows the breakdown as recorded in a study by WRAP (the Waste & Resources Action Programme) from 2002.

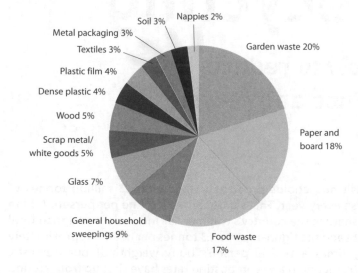

... and what happens to it

Unrecycled waste is either buried or burned. Burying waste at landfill sites is the most common solution in the UK, accounting for more than two thirds of the domestic total.

Landfill

Landfill is a simple approach to waste disposal: put the rubbish in a hole in the ground and cover it with a clay "cap". The main environmental problems of landfill, aside from the potential visual impact and the famously bad smells, are the byproducts. These include leachate – water contaminated with heavy metals and other harmful chemicals – and landfill gas, a mixture of methane, CO_2, vinyl chloride and hydrogen sulphide.

People worry about the plastic being sent to landfill, but it's the biodegradable waste that causes the biggest problem in terms of global warming. Each tonne of degradable waste produces 300–500 cubic metres of landfill gas, much of it in the form of methane, a greenhouse gas more than twenty times as potent as CO_2. Many of the bigger landfill sites extract the methane and use it for power generation. This simultaneously turns the methane into CO_2 – thereby reducing its greenhouse impact – and obviates the need for extra fossil fuel power stations. Where power generation isn't an option, however, the landfill gas seeps into the atmosphere. This leakage accounts for around a third of UK methane emissions.

The proportion of our rubbish heading to landfill is going down. Since 1996, councils have had to pay a landfill tax for each tonne of refuse buried. The tax – and an EU directive to slash the amount of biodegradable municipal waste going to landfill by 2020 – has helped push up recycling rates significantly.

You can find out about landfill sites in your area on the Environment Agency's "What's in my Backyard?" website. If you want to experience one for yourself, the Carymoor Environmental Centre in Castle Cary, Somerset, offers tours of an operational site including the tipping face, composting and recycling operations, leachate treatment and methane electricity generation.

Environment Agency www.environment-agency.gov.uk
Carymoor www.carymoor.org.uk • 01963 350 143

Incineration

The UK incinerates about a tenth of its waste – far less than some other countries such as Denmark and Sweden, which burn 53% and 46% respectively. In most cases, the idea is to combine waste disposal with energy recovery methods such as combined heat and power, which generates electricity and hot water from the burning refuse. (Hospital waste is one of the main sources for incineration and many healthcare trusts now use energy from the incinerators to reduce their heating costs.)

Although incinerators release CO_2, if they're generating electricity or useful heat, they don't add to emissions overall as they reduce the need for fossil fuels. Since they avoid organic waste breaking down into methane in landfill sites, incinerators may even be beneficial in reducing global warming. Despite this, some green groups disapprove of incineration. One reason is that they see it as a disincentive to recycling (once a site is built, a steady stream of waste is required). Another is that incinerators are associated with dangerous dioxin pollution and carcinogenic ash. There are now strict laws governing these poisonous byproducts, but some environmentalists remain unconvinced of incineration's benefits.

1: Reduce

Less in = less out

Recycling and reusing are all well and good, but the very best approach to reducing waste is to generate less in the first place. There are, of course, myriad ways to do this, most of which require little more than a bit of thought. Following are a few good ways to get started.

Food and food packaging

An astonishing third of the food that we buy is fed to the bin, either as scraps or because it's no longer fresh. This waste costs an average household hundreds of pounds per year and results in a considerable amount of methane. The solutions are simple, though they may require a bit of getting used to: planning more carefully when shopping and cooking; using the freezer to prolong the life of things that are likely to go stale; and making use of leftovers rather than letting them sit in the fridge decaying. Finally, try to compost whatever waste food you do create (see p.132). This avoids the creation of methane because the food gets to break down aerobically, in the presence of oxygen. The result is CO_2, a much less powerful greenhouse gas than methane.

For lots of advice on food waste and how to avoid it – including a portion calculator to help you buy the right amount of ingredients – see:

Love Food Hate Waste www.lovefoodhatewaste.com

Food packaging

Of course, it's not just the food itself that fills up our bins: it's the packaging that it comes in. No one knows exactly how much food packaging we throw out each year, but it clearly accounts for a significant proportion of our domestic waste. Perhaps the best strategy here – aside from favouring minimally packaged brands and products – is to follow the advice of the German government and make a point of leaving excess packaging in the supermarket after you've been through the checkout.

Paper

In most people's minds, the environmental problem with paper is the cutting down of trees. In truth, the vast majority of our paper products are sourced from well-managed forests in countries such as Scandinavia, where the demand for paper, if anything, is actually increasing the total area of forest. A much bigger problem than the harvesting of the trees is the energy it takes to produce the paper. This energy causes greenhouse emissions – roughly three kilograms of CO_2 for each kilo of high-quality paper made from virgin materials. Recycled and low-grade paper can be produced with less energy, but it still adds up. The average UK citizen gets through around 200kg of paper products per year if you include everything from notepads and magazines to cereal boxes. The result, according to one breakdown of the typical UK carbon footprint, is slightly more than a fifth of a tonne of CO_2, equivalent to driving a typical car about 600 miles.

Besides cutting back on food packaging, we can reduce our paper use, and the associated emissions, in various ways.

Efficient printing

It's easy to halve your consumption of printer paper by getting into the habit of printing on both sides. This can be automated on some

Despite concerns about trees, the main environmental problem with paper is its embodied energy

high-end printers though in most cases it just means keeping a stack of paper that has been printed on one side and refeeding it into the printer the other way up.

You can also avoid some paper waste by making sure all your documents come out exactly as you intend. This is easily done by hitting the Preview button that appears in the options box after you select Print from the File menu of whichever application you're using.

Finally, bear in mind that you can often substantially reduce the paper needed when printing a document just by reducing the font size, line spacing and margins – or by printing two pages to each sheet of paper, which again can be easily done in the print options box.

Junking the junk mail

If you're fed up with picking junk mail off the doormat and putting it straight into the recycling bin, register with the Mail Preference Service to have your name removed from the lists of direct mail companies. To also stop junk mail sent to "The Occupier", write to Door to Door Opt Outs.

Mail Preference Service www.mpsonline.org.uk • 0845 703 4599
Door to Door Opt Outs Royal Mail, Kingsmead House, Oxpens Rd, Oxford OX1 1RX

Going digital

A final way to save paper is to read news, product information and other written material online rather than buying newspapers, requesting brochures and so on. As we saw in chapter six, browsing websites isn't a carbon neutral activity, but according to most estimates it's still far greener to read online than on paper.

Buying recycled paper products

As well as simply reducing your paper use, it makes sense to favour recycled paper products when possible – not just for kitchen and toilet roll but also printer paper and (for the truly committed) stationery. As discussed on p.129, recycled paper is considerably greener than paper made from virgin materials. For a full range of recycled paper products, visit:

Recycled Paper Supplies www.rps.gn.apc.org

Carrier bags

Few environmental problems get quite as much attention as plastic bags, but are these disposable carriers really such an ecological disaster? Certainly it's easy to see why people object to them: they're ugly, they're rarely recycled, and they take hundreds of years to break down in landfill sites. If they're not disposed of properly, they can also litter streets, beaches and the countryside, and even pose health threats to wild animals.

Measured purely in terms of climate change, however, carrier bags are far less of a problem than many people believe. Figures vary, but no one claims that plastic bags represent more than a twentieth of one percent of the UK's greenhouse emissions. That's despite the fact that for the last few years the UK has got through around 17.5 billion carrier bags per year – almost one per person per day. One reason the impact is so low is that plastic bags are very light and thin, meaning they don't take much energy to manufacture or transport. In addition, because they're not biodegradable, they don't break down into CO_2 and methane – at least not in the current century while we're battling to reduce emissions.

What about the degradable carriers that are becoming increasingly popular at shops keen to flaunt their environmental credentials? The data isn't available to say for sure, but it's likely that these bags – which are made from oil but include special additives to encourage the polythene to break down more quickly – have a slightly bigger climate impact than traditional plastic ones. That's partly because of the extra chemicals they require and partly because they turn into CO_2 within the space of a few years. At least these bags don't break down into methane, however, unlike the plastic replacement packaging made from organic substances such as corn starch and used for sandwich wrappers among other purposes.

Plastic vs paper vs reusable bags

Paper bags tend to be heavier and more energy-intensive to produce than the plastic alternatives, so the carbon footprint of their manufacture is usually higher. In addition, they can quickly break down into methane in landfill sites. On the other hand, paper bags can be easily recycled, composted or even used to generate energy in the form of biogas. For that reason, it's hard to say whether plastic or paper bags are "greener" – it depends on the criteria you choose and on how the various bags are disposed of. (Even plastic bags can be recycled in theory: it's just that few processing facilities exist at present.)

Of course, a carrier that gets reused will typically be greener than one that is disposed of after just one use. And in most cases the greenest option of all will be to avoid carrier bags altogether and bring sturdy reusable bags with you to the shops. ("Most cases" because some bulky but low-quality bags with a lifespan of just a few months may take more energy to make than a few months' worth of plastic carriers.) Reusable bags have become increasingly common for grocery shopping since 2008, when many leading supermarket chains starting charging for the bags they'd previously been giving out for free. It remains to be seen whether this voluntary system will eventually be replaced by an all-out ban on plastic bags, as is being rolled out in the

Paper carrier bags typically have larger carbon footprints than plastic carrier bags do

Indian capital, Delhi. According to rules proposed in early 2009, Delhi's shopkeepers and citizens could face a stiff fine – or even a theoretical prison sentence – for using, storing or selling plastic bags.

Either way, the key point to bear in mind is that carrier bags *aren't* really a key environmental issue – at least not compared to what goes into them. A year's worth of carriers will usually have a carbon footprint smaller than a single beef steak.

Nappies & sanitary products

Babies account for a small percentage of family body mass, but when kitted out with disposable nappies they can easily generate half the contents of a household's bins. A typical baby gets through around five thousand disposables during its nappy days; across the UK, this adds up to eight million per day and three billion each year. Most of these end up in landfill sites where, according to environmental groups, the plastic will take hundreds of years to break down, the super-absorbent granules will soak up groundwater needed for the decomposition of other waste, and the excrement and urine may pose an ecological hazard. Even if these worries are overcautious, as some commentators claim, nappies are energy- and resource-intensive to produce and therefore also expensive to buy. Parents spend an average of £700 per baby on disposables, according to Market Intelligence.

Cloth nappies versus disposables

For all the above reasons, the green advice has usually been to opt for washable cloth nappies – which are also now promoted by local councils keen to cut down on landfill costs. However, a recent in-depth study from the Environment Agency has left many people wondering whether cloth nappies are worth the hassle. The report concluded that although a baby's worth of disposable nappies uses more oil than washables (93kg vs 28kg of crude), it leads to less CO_2 emissions (437kg vs 507kg) and less water usage (34,000 litres vs 86,000 litres). The carbon footprint of reusable nappies was even higher if cleaned via a washing service.

These figures have been disputed, however. The Women's Environmental Network, long-standing proponents of washable nappies, point out that the study made various assumptions that wouldn't apply to most green-minded people. They showed that if you have an energy-efficient washing machine, use a 60-degree wash cycle, limit yourself to 24 nappies, and don't tumble dry or iron them, then the washable option ends up producing a quarter less CO_2 than disposables.

In short, then, cloth nappies will always be better for landfill; whether they're better for climate change depends on how you use them.

Sanitary products

Sanitary towels and tampons raise similar issues to nappies. An alternative popular with some environmentally conscious women is a product called Mooncup. Made from soft silicone rubber, it's a small cup that is worn internally and collects fluid without leakage or odour. It needs emptying less frequently than towels or tampons need replacing. Mooncups cost £18 but last for years and will start saving you money in around six months.

Mooncup www.mooncup.co.uk

2: Reuse
Don't throw it – donate it or sell it

Much of what we throw away is in perfect working order but no longer useful to us. In the UK, we're quite good at donating unwanted books and clothes to charity shops and jumble sales, but many other usable items are simply junked, including some which could easily be donated to others. This is inherently wasteful and also contributes towards the requirement for new items, each of which causes CO_2 emissions in its production.

Following is a list of items which charities will gratefully take off your hands. For items not listed, try Freecycle – a website used by millions of people around the world to give things away to people in their local area. Sign up for your nearest group (it's free) and you can offer anything, with the exception of living creatures, to the other members. If someone wants something that you've offered – and in big cities, especially, the replies come surprisingly thick and fast – then it's up to them to come and collect it. The only problem is that, once you've signed up, reading the offerings of the other Freecyclers can be rather addictive.

Changing the world one gift at a time.

Freecycle www.freecycle.org

Of course, there's a similar environmental case for donating or selling things you would never throw away but which, nonetheless, you're unlikely to actually use. It's possible to sell almost anything via eBay. You'll make money; someone else will save money by buying secondhand; and one less item will need to be manufactured.

eBay www.ebay.co.uk

Donating unwanted items to charity

Bikes

The Re~Cycle project gathers secondhand bicycles and ships them to various African countries. Partner groups teach local people how to repair and maintain them. The organization will accept any type of cycle, as long as they can be put into "going condition" without too many hours of work. For more information, and to find drop-off points, see the following website. Some larger Oxfam shops will also accept bikes.

Re~Cycle www.re-cycle.org

Computers

It's possible to donate computers that are less than around five years old to ComputerAid and Computers for Charities, which will send them to charitable groups in developing countries. You can also offer computers, components and peripherals to other individuals in the UK via Donate a PC.

Computer Aid International

ComputerAid www.computeraid.org • 01323 840 641
Computers for Charities www.computersforcharities.co.uk • 020 7281 0091
Donate a PC www.donateapc.org.uk

Curtains

If you have some high-quality curtains that you no longer want (nothing old accepted), the Curtain Exchange will try to sell them, and donate them to charity if nobody bites.

Curtain Exchange www.thecurtainexchange.net/secondhand.htm

Furniture

There are many charitable organizations across the UK who take unwanted furniture and pass it on at affordable prices to those who most need it. The Furniture Re-use Network (FRN) is the coordinating body for such groups. Because of rules on fireproofing it is worth telephoning a charity before delivering a sofa or armchair.

Furniture Re-use Network www.frn.org.uk • 0117 954 3571

Hearing aids

Help the Aged clean and distribute working hearing aids throughout India. Simply post them to:

HearingAid Appeal Help the Aged, FREEPOST LON13616, London, EC1B 1PS

Paint

Many local-authority recycling centres accept old paint. In order to stop it drying out, cover with cling film and put the lid back on firmly. If the label has been lost write the colour with a marker pen on the can. Alternatively donate it to Community Repaint so it can be used to decorate a community building.

Community Repaint www.communityrepaint.org.uk

Spectacles

More than 200 million people in the developing world would benefit from a pair of spectacles, so consider donating old pairs. Some local authorities accept spectacles as part of their doorstep recycling schemes. Alternatively, give them away via one of the following projects or by dropping them into any Oxfam shop.

Help the Aged World in Sight Appeal www.helptheaged.org.uk • FREEPOST LON 13109, London, N1 9BR • Drop off at Dolland & Aitchison opticians
Vision Aid Overseas Second Sight Project www.secondsightproject.com • Drop off at Kodak Lens Vision Centres

Stamps and coins

Many charities can make money out of old stamps and coins. It is useful to separate your stamps into UK and foreign categories as a kilo of UK stamps will make about £1.50 while a kilo of foreign stamps might make £12.50. The following all accept stamps and coins.

Guide Dogs for the Blind Association www.gdba.org.uk • 0118 983 5555
Oxfam www.oxfam.org.uk • 0870 333 2700
Royal Society for the Protection of Birds www.rspb.org.uk • 01767 680 551

Tools

Tools are badly needed in many developing countries, so if you have unwanted hammers, spanners or pretty well anything else, give them away via one of the following charities:

Tools with a Mission www.twam.co.uk • 01473 210220
Tools for Self Reliance www.tfsr.org • 02380 869 697

Toys

Toys can be donated to the National Association of Toy & Leisure Libraries, who offer them on free loan to parents around the country.

National Association of Toy & Leisure Libraries www.natll.org.uk • 020 7255 4605

Step 3: Recycle
Why it's worth it and how to do more

The rate of household recycling in the UK has doubled in the last few years, but the country still trails far behind some of its neighbours. British homes currently manage an average of 35% (including compostable kitchen and garden waste), even though most experts agree that 60% is perfectly feasible and 80% may even be achievable. The main obstacle – aside from people failing to use the services that already exist – is the expense of collections relative to the current low value of recycled materials. Aluminium cans fetch £500 per tonne but you need 60,000 cans to reach that weight. Recycled plastic bottles can sell for £100–300 per tonne but steel (£60 per tonne) and glass and paper (£25 per tonne) are not exactly money-spinners. The government has set up WRAP (the Waste and Resources Action Programme) to develop markets for recovered materials, but there's still some way to go – a situation that hasn't been at all helped by the economic downturn since 2008.

The low value of recycled materials has led some commentators to suggest that recycling is a waste of time, but such people often ignore two facts. First, recycling almost always saves energy and greenhouse emissions compared to using virgin materials (not to mention cutting down on landfill and reducing the various environmental impacts of raw material extraction). Second, the value of a product is the result of all kinds of factors, including tax structures. If the government cracked down harder on greenhouse emissions, for instance, then the value of recycled materials would shoot up.

That said, it's hard to gauge the full environmental footprint of recycled materials since there's no official record of where they end up. Recycling sceptics point out that around a third of the UK's waste paper and plastic is estimated to be shipped to China, where environmental laws aren't particularly strict. To be fair, however, most of this shipping makes use of

Recycling saints and sinners

The top recyclers in the UK are the residents of St Edmundsbury in Suffolk, who recycle or compost more than 50% of their household waste. In Europe, the greenest region is Flanders in Belgium, which manages a remarkable 71%.

On a national level, the Netherlands tops the EU table, recycling or composting 63% of total municipal waste, followed by Austria (60%) and Germany (60%). The UK, with 35%, is a poorer performer, though still way ahead of the countries at the bottom of the list, such as Portugal (18%).

Globally, New Zealand is perhaps the most progressive country when it comes to waste and recycling and is aiming to minimize "and eventually eliminate" all waste from domestic, construction and demolition sources. For more information, see:

Zero Waste www.zerowaste.co.nz

containers that would otherwise be empty and it's inevitable that materials – raw or recycled – will be in greatest demand in countries with active manufacturing industries. So it's hardly surprising that China takes a large slice of our recycling.

Some recycling in the UK is still done at bottle banks and civic amenity sites, but the number of doorstep collections has expanded widely, and should cover everyone in the country by 2010. If you don't have doorstep recycling, or you want to recycle something too big for your box, locate your nearest recycling bank via:

Recycle Now www.recyclenow.com • 01743 343 403

What can be recycled?

Different local authorities accept different items for recycling, but in some cases there are other ways to recycle those things that the council won't collect. The following pages briefly outline the why and how of recycling for a handful of different materials and items.

Aluminium foil

Aluminium is a highly energy-intensive metal to produce from its raw material, bauxite ore. Recycling cuts this energy demand by a remarkable 95%. For this reason, it's well worth recycling aluminium foil, unless it's particularly dirty. Note, though, that some "foil" – such as the stuff often used in tea bag and crisp packaging – looks like aluminium but isn't. You can tell with the scrunch test: if it springs back it's non-recyclable plastic.

Batteries

Batteries can be environmentally problematic since they contain heavy metals such as cadmium and nickel. Most can easily be recycled and yet more than 90% of used batteries from British homes end up in the trash. If your doorstep collection scheme accepts batteries, then use it; otherwise, store them in a box and drop them off at your local recycling point once in a while.

Also consider replacing dead batteries with rechargeable ones. Modern recharging units are inexpensive and fast-working, and rechargeable batteries have much greater capacity than they used to. For maximum green points, pick up a solar-powered recharger from a specialist outlet such as the Centre for Alternative Technology (www.cat.org.uk).

Cans

Steel is used for most food tins, while aluminium is used for around three quarters of drinks cans. Both metals can be recycled indefinitely, leading to substantial energy and landfill savings (we currently landfill fourteen million dustbins' worth of recyclable aluminium cans each year). So, try to recycle all your tins and cans – including pet-food cans, which many people throw in the bin to avoid washing out. If you don't have doorstep recycling, and you're short of space, a wall-mounted can crusher might help. They're available for £14 from www.recyclenow.com.

If you're really keen, you could consider setting up a can recycling point at work and donating the cans collected to good causes through Novelis Recycling's Cash for Cans initiative:

Cash for Cans www.thinkcans.com

Cooking oil

Used cooking oil can be turned into fuel for diesel vehicles (see p.162), but in most cases it's impractical for fuel processing companies to collect small quantities, such as you might generate at home. One solution is to set up a local collection depot (schools are a good bet, not least because they generate waste cooking oil themselves) where people can drop off used oil and a local biodiesel producer can pick it up. You can find local producers via:

Reuze www.reuze.co.uk/vegoil.shtml

Electrical equipment

In the space of a lifetime, the average British citizen generates 3.3 tonnes of waste electrical and electronic equipment (WEEE) – products such as fridges, computers and TVs. For white goods, most retailers will take away your old machine for recycling or safe disposal when you buy a new one. For smaller items such as PCs try your local recycling centre (and the donation services listed on p.122) before resorting to the bin. Businesses with computers, monitors, printers and cables to recycle should contact:

WEEE Care www.weeecare.com • 01757 708 180

If you'd like to see what 3.3 tonnes of electrical waste looks like, check out the WEEE Man project at:

WEEE Man www.weeeman.org

Engine oil

Used engine oil is highly polluting if not disposed of properly. It's easily recyclable, but you'll need to find your nearest oil bank. That's quickly done with this service from the Environment Agency:

Oil Banks UK www.oilbankline.org.uk • 08708 506 506

Glass

Glass is made from sand, soda ash, limestone and additives for colour and durability. There are no shortages of these materials but recycling saves energy and emissions, as well as reducing the environmental scarring associated with mining. Recycling just two bottles saves enough energy to make five cups of tea, while recycling a tonne saves 315kg of CO_2 emissions.

Brits now recycle more than half their container glass (more than a million tonnes). That's a respectable enough figure but far less impressive than the rates achieved by the Swiss and Finns, who manage 90%. One challenge for the UK is that we import a lot of wine in green bottles and export a lot of whisky in clear bottles. Hence we end up with a lot of recycled glass that isn't much use to our domestic drinks industry. As a result of this, green bottles made in the UK use 85% recycled material, but the clear ones use much less. Excess green glass can, however, be turned into other materials – everything from sand for golf-course bunkers to "glassphalt" for road resurfacing (fourteen million bottles were used to surface the M6 motorway). For more information about the products made from recycled glass see:

WRAP Glass Recycled Products Guides www.recycleglass.co.uk

The business sector has an appalling record on glass recycling. Each year, 600,000 tonnes of bottles and glasses from pubs and clubs ends up in landfill – mainly because green, brown and clear glass gets mixed and it is not economical to sort this prior to recycling.

It's worth noting that even recycled glass bottles have a fairly large carbon footprint, not just because of the heat and power used in their manufacture, but because they're heavy and bulky and therefore energy intensive to transport. Despite popular misconception, from a climate change perspective, a plastic bottle is likely to be considerably greener – especially now that most local authorities accept plastic bottles for recycling.

Mobile phones

There are estimated to be more than a hundred million unwanted mobile phones languishing in British homes. These can be donated to charity – for example, to Oxfam's Bring Bring scheme, which by 2007 had already raised more than £300,000 and stopped 22,500kg of electronic waste from being landfilled. Either drop them off at a local Oxfam store or put them in a jiffy bag and send to:

Oxfam Bring Bring Scheme Freepost LON16281, London, WC1N 3BR

Paper

Despite the occasional urban myth stating the opposite, recycling paper products is *far* better than throwing them away. The exact figures depend on the type of paper being produced and recycled, but making paper from recycled materials typically uses around 64% less energy and 58% less water than producing it from virgin fibres. Put another way, each kilo of recycled paper can save as much as two kilos of CO_2. And that's before you consider the CO_2 and methane that would have been generated if the paper was sent to landfill.

In addition to being used to make new paper, recycled paper can be turned into everything from cat litter and paints to loft insulation (see p.59).

Plastics

Eight percent of the world's current oil production is used to produce plastics, and the resulting products occupy around a quarter of the typical landfill site. Recycling plastic can save 66% of the energy consumed in manufacture, reduce water use by 90% and cut emissions of sulphur dioxide, nitrogen oxides and CO_2 by more than 50% each. Recycled plastic is very versatile and can be turned into products ranging from window frames and sleeping-bag filling to clothes (25 two-litre drinks bottles make one fleece jacket)

Recycling symbols

 The so-called mobius loop simply specifies that an item can be recycled. That doesn't mean that it can or will be recycled in your local area.

 The mobius loop with a percentage symbol in the centre tells you what proportion of an object has been made with recycled materials.

 In certain European countries, the Green Dot symbol is used to show that the producer of a piece of packaging has contributed to the cost of its disposal or recycling. Within the UK it means nothing at all – so don't think items bearing the Green Dot can or will be recycled.

Plastics

Most plastic packaging displays a small symbol specifying the type of plastic used. The following shows the scientific names and common uses for each of them. Numbers 1, 2 and 4 are the most widely recycled in the UK, but check with your local council to find out which they will and won't accept.

PETE
Polyethylene terephthalate: fizzy drink bottles and oven-ready meal trays

HDPE
High-density polyethylene: milk and washing-up liquid bottles

V
Polyvinyl chloride: food trays, cling film, soft drink and shampoo bottles

LDPE
Low-density polyethylene: carrier bags and bin liners

PP
Polypropylene: margarine tubs, microwaveable meal trays

PS
Polystyrene: yoghurt pots, foam, meat and fish trays, hamburger boxes, egg cartons, vending cups, plastic cutlery, protective packaging

OTHER
Melamine: plastic plates and cups

Despite all this, comparatively little plastic is recycled at present. That's due partly to technical difficulties involved in processing large-molecule materials and partly to problems related to separating the various different types of plastic (see box opposite). Most local authorities currently accept plastic bottles only, while larger supermarkets offer plastic bag recycling. Unless your council says otherwise, don't be tempted to put yoghurt pots, margarine tubs, cling film or other plastic packaging in the recycling. If you want your council to do more, writing to your MP (see p.41) is more productive than contaminating the load.

Printer cartridges

Over two million non-biodegradable printer cartridges are sent to UK landfill sites every year. Many of these can be reused or recycled. You can refill inkjet cartridges and save 60% off the price of new ones via services such as:

Cartridge World www.cartridgeworld.co.uk • 0800 183 3800

Laser printer toner cartridges can be donated to various charities who can make money from recycling them. These include:

Action Aid www.actionaidrecycling.org.uk
Rain Forest Concern www.rainforestconcern.org

Tetra Pak cartons

Most milk and fruit juice that doesn't come in plastic bottles comes in Tetra Pak cartons. These are most commonly made of 74% paper, 22% polythene and 4% aluminium film. The fibre within them is quite valuable and can be recycled into new paper products that require strength. However, you'll probably find that your local authority doesn't accept them for recycling.

If you're determined to recycle as much as possible, check the following website to see if there's a carton recycling point near you. If there isn't, the best strategy is simply to avoid buying Tetra Paks where possible.

Tetra Pak Recycling www.tetrapakrecycling.co.uk

Tyres

In the UK, 28 million tyres are discarded each year but since 2006 their disposal hasn't been permitted in landfill sites. As they have an energy content of 32 gigajoules per tonne there is significant economic potential for tyre incineration. But it makes no sense to burn a tyre until it's no longer functional. Virtually all tyres can be retreaded several times. Each retread prevents the manufacture of one new tyre, saving twenty litres of oil for a car tyre and 68 litres for a truck tyre. Find out more about retreads from:

Retread Manufacturers Association www.retreaders.org.uk • 01270 561 014

Find out more

For more information about waste, recycling and related issues, drop in to the following websites:

Recycle Now www.recyclenow.com
Recycle More www.recycle-more.co.uk
Waste Online www.wasteonline.org.uk

For advice on minimizing waste and reducing environmental impact from a business perspective, try:

Envirowise www.envirowise.gov.uk • 0800 585 794

Of course, in addition to recycling more, it also makes sense to purchase recycled products wherever possible. This way you can boost the market for recycled materials and help to "close the loop". To find local suppliers of recycled products ranging from toilet tissue and wrapping paper to plant pots and kitchen cupboards, visit:

Recycled Products Guide www.recycledproducts.org.uk

Composting
Organic recycling

Over 50% of our domestic refuse is organic in nature. That includes not only paper and cardboard – which can be easily recycled – but garden cuttings and kitchen scraps that need instead to be composted (or possibly turned into useful energy in an anaerobic digester). Many local authorities now offer doorstep collection schemes for this waste or have specially designated skips at the local tip. However, if you already have your own compost heap, keep up the good work: this is still the most ecologically friendly way to treat your food waste and garden clippings. As long as the heap can "breathe" it will not produce the greenhouse gas methane.

Instead the heap will produce CO_2 – but less than was absorbed from the atmosphere by the decomposing trees and plants as they grew. With a bit of patience you'll end up with a high-quality crumbly compost that will help your garden thrive.

How to make compost

There are many schools of thought on composting, but the crucial thing is to get good results without creating bad smells or attracting vermin (especially important if you only have a small garden and there are children about). The best bet is to buy a sealable compost bin, which many local authorities will sell you at a highly subsidized rate. These are mostly made from thick plastic and will take virtually any organic matter. If you build your own compost bin on bare soil and leave it open at the top, you should avoid all foodstuffs except fruit, vegetables and things like tea leaves and coffee grounds (meat, dairy and grain-based products are a magnet for rodents).

Whichever option you go for, ensure you have a good mix of carbon-rich browns – for example dried flowers, woody stems and cardboard – and nitrogen-rich greens, such as fresh grass cuttings and kitchen waste. Best results are achieved by cutting large items into smaller pieces to accelerate the process. Build the heap in layers (rather than just piling it up) and introduce air on a regular basis by occasional turning or by adding a layer of cardboard and paper every now and then. You want a moist heap rather than a wet or dry one; so if it's too wet add more dry material and if too dry add water, or – if you can face it – urine, which is a great accelerator as it's very rich in nitrogen. Other good accelerators are young nettles and comfrey leaves. Your heap will take between six months and two years to turn into sweet-smelling dark crumbly compost.

A can-o-worms wormery from Wiggly Wigglers

For the intrepid a wormery is another option. These come in all shapes and sizes (so can be good for small spaces) but they demand rather more care and attention than a normal compost heap. For more information, or to buy online, visit:

Original Organics www.originalorganics.co.uk
Wiggly Wigglers www.wigglywigglers.co.uk

If your garden has plenty of deciduous trees, it's well worth gathering the fallen leaves in autumn and storing them either in a separate heap or – if there aren't that many – in a black bin liner. By the following year this will have produced a nice crumbly mould which can be used as a mulch or to enhance the texture of your soil (evergreen leaves take much longer).

If you're feeling really adventurous, you might even consider recycling your own human waste via a composting toilet. There are a surprisingly large number on the market. For information and further links, go to:

Composting Toilet World www.compostingtoilet.org

For more information on composting in general, visit:

Composting Association www.compost.org.uk
Recycle Now www.recyclenow.com/home_composting

Water

How to use less

Mention green living and the first thing that comes to some people's minds is minimizing water wastage – a brick in the toilet cistern, perhaps, or turning off the tap while brushing teeth. That's understandable when you consider that water availability is one of the most pressing global issues. But how much does saving water really matter in regions where there's plenty?

Most eco-minded articles on water fudge the answer to this question by focusing on the terrible inequality in access to water between rich and poor countries. It's true that this inequality is extreme: UK citizens use around 55,000 litres of water every year, while in other parts of the world more than a billion people lack access to clean water and 21 million people die each year from diarrhoeal diseases caused in part by poor sanitation. It's also true that the outlook is dire: the proportion of people living in water-stressed areas is expected to grow from one third to two thirds over the next 25 years. Appalling though these facts are, no one is suggesting that we ship our spare water abroad, so it's hard to see how any of this is relevant to our own water use.

A more pressing reason for saving water is that we might one day not have enough ourselves. Hosepipe bans in southeast England have shown how a combination of dry winters, warm summers and leaky infrastructure can leave us with only just enough to go around. Add in a growing population in certain regions and increasing numbers of people living on their own (a trend which is boosting the number of water-hungry appliances), and it's clear that there could be water stresses in the future. Water companies are already being forced to extract more from underground reservoirs, which are not being replenished, and a desalination plant is currently being built on the Thames, at Beckton in the Borough of Newham.

Another reason to cut back on unnecessary water consumption is that every drop we consume has been treated and pumped; and every drop we put down the plughole or toilet is later treated and pumped again. All of this treatment and pumping requires energy, and therefore results in emissions of greenhouse gas. The UK water industry uses around 6000 gigawatt hours of energy to provide water and treat sewage each year, producing 2.7 million tonnes of CO_2 – around 0.5% of total UK emissions. (When the desalination plant is finished, this total might shoot up by around 10%.) To put these figures in perspective, a typical British family of four uses approximately 200,000 litres per year. This requires 94kWh to supply and 83kWh to treat, resulting in 78kg of CO_2. That's equivalent to an average car travelling 1100 miles.

With all this in mind, the Centre for Alternative Technology suggests we should aim to reduce our daily water use to about eighty litres per person. This chapter suggests some tips for doing just that.

Understanding your water use

The average British resident currently uses approximately 150 litres of water per day. As the diagram shows, around a third of this goes towards toilet flushing, and another quarter is used for washing. Of course, these proportions vary widely from home to home. If you regularly water your garden with a hose or sprinkler, for example, then that may well constitute the majority of your consumption.

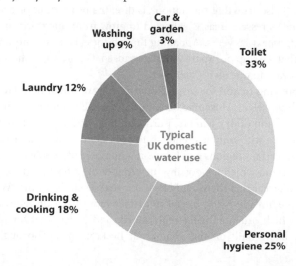

Washing up 9%
Car & garden 3%
Toilet 33%
Laundry 12%
Typical UK domestic water use
Drinking & cooking 18%
Personal hygiene 25%

Water metering

At present, only a minority of UK homes (approximately a third) have a metered water supply. The rest are billed a fixed amount based on expected use. If you're careful with your water consumption, however, you'd probably be well advised to request water metering, something available for free to all homes. This way you only pay for what you use. If you've reduced your consumption to eighty litres per day, as the Centre for Alternative Technology recommends, a meter might take 40% off your bill. Even if you haven't yet managed to reduce your water use, the evidence suggests that switching to a meter might help you do so. On a metered tariff, the typical home reduces its water consumption by 5–10% as the residents get used to the idea that there's a cost attached to each litre.

Hot water versus cold water

Though it makes sense to limit all forms of water waste, in a relatively wet country such as the UK the key issue in homes is not water use in itself, but the use of hot water. As we saw on p.50, water heating accounts for around 1.25 tonnes of CO_2 emissions in many British homes. If we assume that a third of the water we use comes from hot taps, and the remaining two thirds is cold, then a back-of-the-envelope calculation would suggest that each litre of hot water we use has a carbon footprint around a hundred times greater than a litre of cold. So while saving cold water is important, saving hot water is much more important still.

Virtual water

The water we consume directly at home constitutes only a minority of our total water footprint. In just the same way that most items we buy come with some embedded energy, many items also have embedded water – or "virtual water" in the green jargon. This is especially true for food products, which often take a huge amount of rainwater or irrigation to grow. According to a 2008 study by WWF, UK citizens consume around thirty times more water virtually – in food and textiles products – than they do directly. This makes the average UK water footprint a massive 4645 litres a day. Of this, around 62% comes from abroad – the sixth highest figure in the world after Brazil, Mexico, Japan, China and Italy. Interestingly, the WWF study found that the water footprint of a diet containing lots of meat and dairy is as much as twice as high as a vegetarian or vegan diet. As water scarcity becomes an increasingly urgent global problem, this is likely to become a frequently heard argument against meat and dairy consumption.

How to reduce household water use

Toilet flushing

Cisterns that predate 1991 tend to use around nine litres of water, while later models typically use around 7.5 and those installed since 2001 tend to use six. Aside from flushing less frequently (as per the old saying "If it's yellow …"), the easiest way to make savings is to put a displacement device in the cistern. A water-filled plastic bottle will do the trick, if it fits, but you might be better with a built-for-the-job model such as those available from:

Hippo Watersaver www.hippo-the-watersaver.co.uk

The only disadvantage of this approach is that some old-fashioned toilets don't work as well with a displacement device in place. If you find that your flush isn't working as effectively as it used to, try moving the displacement device to a different part of the cistern, if possible, or swapping it for something smaller.

When fitting a new toilet opt for an eco-flush model that allows you to choose a full or half setting. These are widely available, though you might want to turn to a specialist supplier for a super-efficient toilet that uses as little as four litres of water for a full flush and two for a half flush. Stockists include:

Green Building Store www.greenbuildingstore.co.uk

For the really committed, you could consider a reed-bed sewage treatment set-up or even do away with a conventional toilet altogether and use a composting model (see p.119).

In terms of the impact on the environment more broadly, also think about *what* you flush. It only takes a blockage at the sewage works for such items such as tampons and condoms to back up into our waterways and end up in rivers and on beaches.

Baths and showers

A five-minute shower typically uses around 25 litres of warm or hot water, compared to 80 litres for a bath. So there's a good case for heeding the green cliché of taking fewer baths and more showers (as long as you don't have a pressurized power shower, as these can use up to 120 litres a go). Other possible steps include reducing the temperature of the water or taking shorter showers, though you'll get more gain and less pain by opting for an aerated shower head. These reduce the amount of water needed for an invigorating shower by up to 60% by mixing air with the water. Manufacturers include:

Oxygenics www.low-e.co.uk
Bricor www.bricor.com

The American company Evolve has even produced a shower head designed to avoid the water and energy wasted when the water is left running to warm up. A thermostat automatically reduces the flow once the water has reached the correct temperature. When the person gets into the shower they flick a switch and the water starts flowing again.

Aerated shower heads, such as this one by Oxygenics, are popular in hotels, where hot water bills are large, but can be just as effective in homes

Evolve evolveshowerheads.com

If you're sufficiently motivated, it's even possible to reuse bath water for your garden. This can be done with a Drought Buster syphon pump, which runs from the bath to an outside water butt, via a window. You squeeze the device to get the flow going, and gravity should take care of the rest. As well as saving water, this can be very handy during dry spells with hosepipe bans.

Drought Buster www.droughtbuster.co.uk

Watering the garden

Watering the garden with a hose can consume more than one thousand litres per hour, so it makes sense to use alternative sources as much as possible. The first thing to do is get a water butt fitted to the down pipe from your roof and start collecting rainwater. Plants actually grow better in natural water sources so this makes sense from every perspective. Combine a water butt with a solar-powered irrigation system and the garden will even water itself.

If, on the other hand, you're using tap water in your hose or watering can, it's possible to minimize evaporation by watering the garden in the evening and aiming at the base of plants rather than the leaves.

Taps

You can reduce the water consumption of taps with Tapmagic inserts (pictured), which provide a spray when the tap is turned slightly but a full flow when turned completely. They're less than £5 each and are easy to fit yourself.

Tap Magic www.tapmagic.co.uk

Tapmagic inserts can cut water use for a particular sink by as much as 50%

Recycling and harvesting water

Grey water is water that has been used in baths, washing machines and sinks. With a bit of careful plumbing, it can be stored in a tank and recycled for toilet flushing. A more complex system, such as those available from Free Water UK, will even filter the water sufficiently well to make it usable for cleaning.

Free Water UK www.freewateruk.co.uk

Even more serious is a rainwater harvesting system. You'll still need a mains supply for your drinking water but will save around 50% on your water meter. Such systems cost around £2000–3000 for a typical home and pay for themselves in around ten to fifteen years (this will fall if water bills keep rising in coming years, as expected). There are currently only a few thousand water harvesting systems in UK homes, but across Europe around 100,000 are being installed each year. To find out more, see:

UK Rainwater Harvesting Association www.ukrha.org
Rainharvesting Systems www.rainharvesting.co.uk

Part III

Travel

Travel: the basics
Cars & driving
Air travel & alternatives

Travel: the basics

Understanding travel's climate impact

Both at home and abroad, we travel further today than ever before. Each year the typical British resident travels around 12,000 miles within the UK, which marks an increase of more than 50% in just three decades. The vast majority of this domestic transport – around four fifths – takes the form of car journeys. (In the last few years this proportion has started to fall slightly, but the overall distance being driven keeps on rising.) As the price of air travel has dropped, we also leave the country with growing frequency. The British take foreign trips around four times more often than they did in 1980, and more than eighty percent of these trips involve flying.

Around the world, the rise in passenger miles is even more striking. Car ownership and use is rocketing in many developing countries – China, especially – and budget airlines are providing a rapid boost to regional air travel in areas such as South Asia and Latin America. All this extra travel reflects increasing wealth and mobility, but the inevitable down side is a rise in carbon dioxide emissions.

The climate impact of travel

The global transport sector was estimated to be responsible for around 13% of man-made greenhouse gas emissions in 2004 and it seems highly probable that the figure will have crept up since then. Roughly speaking, around half of these worldwide transport emissions are caused by private vehicles – cars, vans and motorbikes – and the other half by public and freight transport, such as trucks, ships, buses and planes.

In the UK, a developed country with a high level of mobility, transport is a relatively bigger source of global warming gas. Collectively, all forms of transport produce 22% of the nation's total greenhouse emissions. And the true total is higher still because, as we'll see in chapter twelve, the warming impact of air travel is bigger than the official figures reveal.

All these statistics provide interesting context, but they don't shed much light on how we can each reduce the CO_2 emissions caused by our travel. For that we first need to consider how the various modes of transport compare.

What's the greenest way to go?

It would be helpful to be able to list all forms of transport and their associated emissions. But comparing the climate impact of different modes of transport is complex.

The main problem is that the emissions of individual vehicles, rather than just the modes of transport themselves, vary widely. A diesel train is typically far more polluting than an electric train, just as a petrol Hummer is much worse than an electric Mini. As we'll see in the next chapter, even the speed that the vehicle is driving at can make a difference to its fuel use. So it doesn't necessarily make sense to talk about the emissions of simply "driving" or "taking the train".

Another challenge is the fact that the only meaningful way to compare the various forms of transport is to look at their emissions per passenger mile – that is, the total emissions of the car, train, plane or bus divided by the number of people on board. This makes broad-brush comparisons difficult because occupancy rates vary so widely. A train on a busy commuter line may consistently have more passengers than seats; a train on a less popular route might have only a couple of passengers per carriage. Moreover, emissions per passenger mile isn't a perfect metric. If you're about to drive to the shops and a family member decides to come along for the drive to keep you company, then the emissions per person will be instantly cut in half. But the total amount of CO_2 emitted will be unchanged (or, if anything, will be fractionally higher due to the extra weight in the car).

A further issue to consider is the argument that a plane, train or bus is scheduled to run anyway, regardless of whether you buy a ticket – unlike

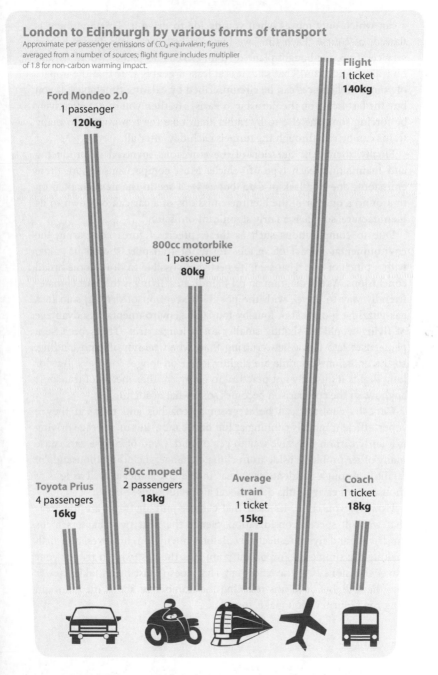

London to Edinburgh by various forms of transport

Approximate per passenger emissions of CO_2 equivalent; figures averaged from a number of sources; flight figure includes multiplier of 1.8 for non-carbon warming impact.

Flight
1 ticket
140kg

Ford Mondeo 2.0
1 passenger
120kg

800cc motorbike
1 passenger
80kg

Toyota Prius
4 passengers
16kg

50cc moped
2 passengers
18kg

Average train
1 ticket
15kg

Coach
1 ticket
18kg

a car, which only moves when you decide to drive it. This argument is flawed, of course. Each time we buy a ticket, we create demand for a service; if no one bought plane or train tickets, no planes would fly and no trains would run. That said, it is at least arguably true that the impact of your travel choices can be circumscribed by existing timetables. If you buy the last ticket on the Eurostar to Paris, say, then you might effectively be forcing someone else to fly rather than take the train. Only so many trains can be run through the tunnels each day, after all.

Finally, there's the question of the emissions involved in producing and maintaining each type of vehicle. Most comparisons ignore these emissions, due to a lack of data, but as we'll see in the next chapter, up to around a quarter of the lifetime emissions of a car can be down to its manufacture, so this is a fairly significant omission.

Due to complications such as these, precise figures comparing the environmental impact of various forms of transport should be taken with a pinch of salt. That said, it's perfectly possible to draw some broad conclusions. As the diagram on p.145 indicates, flying is the least climate-friendly way to travel, with the possible exception of driving solo in a gas-guzzling sports car or four-by-four. (The environmental disadvantage of flying would be slightly smaller for a longer trip. That's because a plane uses less fuel when cruising than when taking off and landing, so the emissions per mile are slightly lower on longer journeys. But for long flights it's usually not practical to go by another means of transport anyhow, so the comparison becomes somewhat academic.)

Cars, by contrast, can be as green as coaches and trains if they're super-efficient models running at full occupancy. But on average driving is a fairly carbon intensive way to get around. (And of course cars cause many other problems aside from climate change, including thousands of fatalities through accidents each year in the UK alone, as well as tens of thousands of early deaths due to local air pollution; see p.149.)

However you cut it, it's clear that car use and air travel are the areas that warrant special consideration. Hence the next two chapters focus on driving and flying respectively. Before moving on, however, it's worth making the simple but important point that the best way to reduce your travel emissions is to travel less far. That doesn't just mean taking fewer long-haul flights; it means reducing the number of short, medium and long road journeys you make, too.

Cars & driving

Lower-carbon motoring

Our contribution to global warming is rarely as visible and direct as it is with driving. At filling stations, we purchase up to fifty litres of fossil fuel at a time. This fuel is burned in the engine, and during the combustion the carbon in the fuel is converted into carbon dioxide, which is subsequently released into the atmosphere. Thankfully, the efficiency of car engines is increasing. Unfortunately, though, the efficiency gains so far haven't been substantial enough to counteract the ever-growing distances driven each year. As a result, CO_2 emissions from cars are still increasing in the UK, just as they are in the world as a whole. Though super-efficient electric cars are starting to appear, the typical car on the road still produces its own weight in CO_2 for every 4500 miles driven. (To put that figure into perspective, a typical car's weight in CO_2 would fill a balloon approximately the size of five double-decker buses.)

This chapter provides information and advice on greener driving, greener cars and greener fuels. First, though, here's some background to help you gauge the environmental impact of your current vehicle.

Understanding your car's carbon footprint

The typical car in the UK travels around 14,500 kilometres per year and emits 181 grams of CO_2 per each kilometre travelled. That adds up to an annual carbon footprint of around 2.6 tonnes – between a fifth and a seventh of the typical British citizen's total, depending on how you work it out. Of course, your own driving footprint could be very much bigger or very much smaller, depending on the efficiency of your car and the distance you drive each year.

To quickly get a rough sense of how green or ungreen your current car is, first consider its size. As basic physics dictates, the bigger and heavier a vehicle is, the more energy it will take to move it around. Next consider the engine. Roughly speaking, the larger the capacity and the greater the performance, the higher the fuel consumption. Finally, consider the age of the car. Older engines tend to be less efficient than newer ones. (For a more precise indication of your car's efficiency, look it up at www.vcacarfueldata.org.uk.)

Regardless of the vehicle that you drive and how far you drive it, it's easy to work out how much CO_2 is released from its exhaust pipe. That's

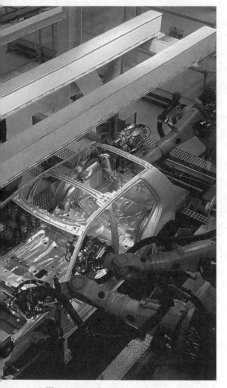

The materials and machinery required to manufacture a car create as much CO_2 as three years' worth of driving

because there's a roughly fixed amount of carbon in each litre of fuel, and practically every gram of that carbon is converted into CO_2 in the engine. According to government-approved figures, each litre of petrol becomes 2.315kg of CO_2, and each litre of diesel become 2.630kg.

In reality, however, the total emissions caused by your car's fuel is greater than the emissions exiting the tailpipe. That's because each litre of petrol or diesel has been extracted, refined, processed and transported before being made available in a filling station. Each of these stages requires energy and causes emissions, but the official figures allocate these emissions to the countries and companies producing and delivering the fuel – not to the driver who ultimately burns it.

More significant than the production of the fuel is the production of the vehicles that consume it. Each car contains hundreds or even thousands of kilograms of metal and other materials, the production of which requires a huge amount of energy even before the components are produced and assembled. According to one expert at the Stockholm Environment Institute, producing a typical car results in a massive eight tonnes of CO_2 – equivalent to more than three years of fuel consumption for the average driver.

Unfortunately, it's pretty well impossible to know the emissions involved in producing any specific car – either one you already own or one you're thinking of buying. You can get an approximate sense by considering the weight; the bigger and heavier the car, the more materials and energy it will have taken to build. But accurate figures are very hard to come by, and partly for this reason it's usually not possible to work out the precise carbon footprint of your driving.

According to the best available models the emissions of the "total" use of a car – including producing, fuelling and maintaining it – are typically around twice the direct emissions of the car itself. By this rule of thumb, the average car is responsible for around five tonnes of CO_2 per year – enough to fill a medium-sized sports hall.

As we'll see later in this chapter, the high level of emissions involved in producing a new car can make it very tricky to work out whether it's environmentally beneficial to upgrade from your current car to a greener model.

Beyond climate change

Climate change may be the most pressing environmental problem associated with our driving, but it's not the only one. Besides globe-warming CO_2, car exhaust emissions also include a cocktail of local air pollutants, including poisonous gases such as carbon monoxide, and particulates – tiny solid hydrocarbon particles that can lead to lung and heart disease, among other health problems.

According to the government, vehicles are largely responsible for the airborne pollution that causes around 25,000 premature deaths and as many hospitalizations each year in the UK alone: almost ten times as many deaths as are caused by road accidents. In countries with older, lower-quality vehicles, the local health impact of exhaust fumes is even greater.

Engines are much cleaner than they used to be but, with the exception of electric and hydrogen vehicles, all cars still contribute some level of local air pollution. As we'll see later, diesels and motorbikes – while being better on average in terms of climate change – are often worse in terms of local pollutants.

Lower-carbon driving

How to reduce the emissions of your current car

Even if your vehicle's a gas-guzzler, you could probably cut its fuel usage by a tenth – and in some cases by as much as a third – by simply tweaking your driving practices. Here's how.

Limit your speed

If you travel much on motorways, then the single most significant way to reduce the carbon footprint of your driving is to limit your speed. Most cars achieve maximum fuel efficiency at around 45–50mph. Above that level, fuel consumption rises by as much as 15% for every additional 10mph. So simply driving on the motorway at 60mph rather than 80mph can cut emissions and fuel costs by as much as a quarter.

(Of course, governments could easily force us to drive in a less polluting way by simply reducing and imposing the speed limits of main roads. This is exactly what happened in the US in 1974, when a 55mph speed limit was rolled out across the country in response to the 1973 oil crisis. To date, though, few if any governments have implemented new speed restrictions in light of climate change.)

In some cases, high speeds can be avoided at no time cost by favouring more direct routes on A-roads over longer routes on motorways.

Drive smoothly

Rapid acceleration and sudden braking will increase emissions both directly, by increasing your fuel consumption, and indirectly, by reducing the longevity of your vehicle and bringing forward the date that a new one will need to be manufactured. So it makes sense to accelerate gradually and also to take your foot off the gas as early as possible when you're approaching a stop. Every time you press the brakes, you're turning energy from the fuel into wasted heat.

The rules are a bit different for hybrid cars (see p.156), which often operate most efficiently at 30–40mph. For many hybrids, it can increase efficiency to accelerate fairly briskly until you reach this speed range and then to employ so-called "pulse and glide" driving – basically, hovering in the optimal speed range through small accelerations and decelerations. When it's time to slow down, brake slowly at first, then increase the pressure: this ensures that the maximum energy goes into recharging the battery versus creating unusable heat.

Whatever car you're in, avoid pushing it too hard before changing gear. If you have a rev counter, try to shift up a gear before you reach 2500rpm in a petrol car, or 2000rpm in a diesel model.

Other ways to shave your car emissions

Keep heavy items out of the car unless you need them – you'll typically lose a percent or two in efficiency for every extra 50kg you haul. Avoid unnecessary roof racks, too, as they increase aerodynamic resistance. Also keep an eye on tyre pressure: rolling resistance goes up and efficiency goes down by as much as 1% for every PSI (pound per square inch) below the recommended pressure range. Never over-inflate your tyres, though, as this increases the risk of accidents while doing little if anything to boost efficiency.

Breakdown organizations

If you use a car but have misgivings about it – for example, you feel that public transport should be prioritized over the needs of drivers – then consider which company you use for roadside rescue. The AA (part of energy group Centrica) and the RAC (part of the RAC Group, which also includes Hyundai cars and a giant vehicle leasing firm) may not be the most disagreeable companies in the world, but they do have a long history of lobbying for road expansion and drivers' rights through groups like the British Road Federation.

The ETA (Environmental Transport Association) was set up specifically as an environmentally sound alternative to the big roadside rescue companies. It offers a similar service at a similar cost, boasting a 35-minute average callout time, an 80% success rate in fixing cars by the roadside, and very high customer satisfaction. While it's not anti-car as such, it "conducts and commissions research into environmental transport issues and lobbies the government to encourage its support of alternatives to the car". It even organizes Green Transport Week and Car Free Day, and also offers roadside rescue for cyclists.

Environmental Transport Association www.eta.co.uk • 0800 212 810

Another fuel drain is air conditioning, which typically cuts down a vehicle's efficiency by a few percent. That said, if it's a choice between driving with the windows down and running the A/C, there may be little difference – at least not at high speeds. That's because wide-open windows will typically increase the car's aerodynamic drag, and this has a higher impact as the speed increases. If outside temperatures are comfortable, try using the vents and fan but leaving the A/C off.

Finally, except when it's required – such as in stop-and-go traffic – avoid idling. Letting the car tick over for anything more than around ten seconds will use more fuel than turning the engine off and back on. In some cars just five minutes of idling can throw half a kilo of CO_2 into the air.

Consider your fuel

As well as changing your driving practices, you may also be able to improve the environmental performance of your car by switching to an alternative fuel, such as LPG for petrol cars or locally produced biodiesel for diesel cars. For more information on these lower-carbon fuels, flick ahead to p.162.

Buying a greener car
The if and how of purchasing a low-carbon vehicle

On the whole, cars are gradually getting greener. This applies both to their fuel efficiency (and therefore the carbon footprint of each mile driven) and to their generation of poisonous gases and particulates. However, some cars are *far* greener than others, so if you're in the market for a new vehicle, it makes sense to seek out the lowest-emissions model that fits your needs and budget. As we'll see, that's easily done. What's harder is working out whether upgrading to a greener car will actually bring about an overall environmental benefit.

Should you buy a greener car?

With all the headlines about eco-friendly hybrids and electric cars, there's no wonder that many people are keen to do their bit and buy a greener vehicle. But if, as we've seen, each new car causes many tonnes of CO_2 to enter the atmosphere just by being produced, will these manufacturing emissions outweigh the benefits the car brings in reduced fuel use? Might it be better to stick with your existing car and to keep it on the road as long as possible?

The answer to this question depends on a huge range of factors: not just the green credentials of the new vehicle, but other things such as the amount you drive each year and what happens to your existing car after you buy a new one. For example, if you currently have a fairly polluting car but you only drive, say, a thousand miles per year, then a shiny new hybrid might take thirty years to save as much CO_2 as was produced in its manufacture. Moreover, if you sell your old car and it gets purchased by someone who drives far more each month than you do, then the equation looks even less eco-friendly: the inefficient vehicle will be doing *more* miles each year than at present, and the new one may be ready for the scrap heap before it's repaid its manufacturing emissions.

On the other hand, if you drive 10,000 miles each year and you're in a position to scrap your existing, more polluting vehicle – or to give or sell it to someone who drives very little – then upgrading makes perfect sense. The emissions caused by producing the new car will be paid off in a few years, and the inefficient older car will be driven fewer miles each year. Moreover, by purchasing new, "at the top of the market", you'll be helping to establish a wider market for greener cars.

Whether it makes environmental sense to trade up to a more efficient car also depends to a significant extent on the model you choose. The smaller and lighter the car, and the lower its emissions in use, the more quickly it will make sufficient fuel savings to counteract the energy used in its manufacture. In other words, it might very well make good sense to trade in your current car for a small electric model; but the calculation would look less good if you were buying a family-sized hybrid.

As already mentioned, one alternative to swapping your existing car is to try to run it on a lower-carbon fuel such as LPG or recycled biodiesel (see p.162).

Green cars: the basics

If you do decide it makes sense to buy a greener car, the main options are an efficient "conventional" car, an electric hybrid, a fully electric model, or a car pre-converted to run on LPG. Electric cars are almost always the greenest, even if charged up by electricity produced mainly from coal and gas. They're still not widely available at the time of writing, though that is likely to change during 2010 and 2011. As for other types of cars, you can check the exact emissions of each vehicle you're considering at the following websites. These list figures for CO_2 and local air pollutants for practically every model from every manufacturer. The ETA site focuses on cars currently on the market and provides a handy overall environmental star-rating for each car, whereas the VCA site has data for second-hand as well as new models. WhatGreenCar offers all sorts of reviews, articles and advice.

ETA Car Buyer's Guide www.eta.co.uk
VCA CarFuelData www.vcacarfueldata.org.uk
WhatGreenCar www.whatgreencar.com

Financial benefits of green cars

The UK government used to operate a scheme called PowerShift, which offered grants to help individuals buy low-carbon cars or convert their current vehicles to run on greener fuels. This, unfortunately, has come to an end, but buying a greener car still can be very attractive financially. The average car on sale at the time of writing will consume approximately £13,000 of fuel by the time it's done 100,000 miles – or maybe more if fuel prices keep rising as expected. Switching to a smaller, greener car could cut that figure by up to £6000. Furthermore, road tax in the UK is now tiered according to emissions, so while some SUVs cost £400 per year just to keep on the road, a car that emits less than 100g of CO_2 per km pays nothing at all. If you live in a city, you might also find that a low-emissions car qualifies you for a discounted or free residents parking permit. In London, electric cars and those powered by alternative fuels benefit from exemption from the congestion charge, too.

New versus second-hand

Is it more environmentally friendly to buy a new low-emissions car or to opt for a second-hand, perhaps less technologically advanced, model? Once again, there's no simple answer to this question. As we've seen, new cars cause emissions in production, but at the same time the best way to influence the overall market is by purchasing a new model.

If you do buy second-hand, the greenest choice of all is probably to pick up an old but reasonably efficient banger that not many other people are likely to want and to keep it running as long as you can. This way you'll be helping reduce the need for new cars while also making sure you use as little fuel as possible. This option won't suit everyone, of course, as it may be a headache in terms of maintenance.

If you're buying a second-hand car for only occasional, low-mileage use, then it arguably doesn't matter quite so much what you choose. Even if you pick a relatively high emissions model, you'll at least be stopping someone else buying it and driving it longer distances.

Petrol, diesel and LPG cars

The least expensive way to get a greener car is to opt for a regular petrol or diesel car that happens to be very light and fuel-efficient. Both fuels have their relative pros and cons. Diesel cars are considerably better than petrol models when it comes to carbon emissions – typically by around 15–20%. On the other hand, diesel engines tend to produce a higher number of particulates and other local air pollutants, so they're perhaps somewhat antisocial for use in urban areas. However, this disadvantage is diminishing as diesel engines get better at limiting exhaust pollution and as growing numbers of filling stations offer cleaner "city diesel" fuel.

When it comes to cost, diesel cars are cheaper to run but also more expensive to buy. The fuel savings will typically have repaid the higher initial price tag after around 50,000 miles. But the carbon savings will kick in right away, and over the life of the car the cash savings will be substantial – especially if oil prices rise.

Good cars, bad cars

Following are the greenest and least green cars (excluding electric models) available in the UK at the time of writing, as determined by the ETA. As the lists show, the most eco-friendly cars are light and have comparatively small engines. The most climate-frying and air-polluting cars tend to be heavy and un-aerodynamic or super-powerful – either boy-racer vehicles or hulking 4x4s.

The ten best...	...and the ten worst
Toyota Yaris	Dodge SRT10
Honda Civic Hybrid	Lamborghini Murcielago
Toyota Prius	Ferrari 599 GTB Fiorano
Renault Modus	Ferrari 612 Scaglietti
Citroën C1	Bentley Motors Arnage
Peugeot 107	Bentley Motors Azure
Toyota Aygo	Bentley Motors Continental
Renault Clio	Ferrari F430
Toyota Auris	Ferrari F430 Spider
Suzuki Swift	Aston Martin Lagonda DBS

One other consideration is the alternative fuels which petrol and diesel engines leave you the option of using. Petrol cars can be converted to LPG fuel, which gives them CO_2 emissions as low as those of a diesel car but with far lower levels of poisonous fumes. For more information on LPG, see p.164. Similarly, diesel engines can be run on biodiesel, though this could increase or decrease your carbon emissions, depending on the source of the fuel (see p.162).

One final thing to look out for when shopping around is a start-stop system. These are currently only widespread on hybrid cars but are starting to appear in regular petrol cars. The device automatically turns off the engine when the car is stationary, and fires it up again instantly the moment you step on the gas. This can help reduce fuel use by up to 10%, with the best savings made by those drivers who spend a lot of time idling in stationary traffic.

Cars with electric motors

Electric motors powered by batteries are inherently capable of higher fuel efficiencies than those achieved by internal combustion engines. That's because more than half of the energy stored in a litre of petrol or diesel gets lost as heat when the fuel is burned in a combustion engine. Electric motors, by contrast, can turn the majority of the energy stored in a battery into movement.

There are three kinds of cars with electric motors: electric hybrids, which have motors powered by petrol; fully electric vehicles that are charged up exclusively by mains electricity; and plug-in hybrids that can be plugged in *and* filled up with petrol. Let's take a brief look at each.

Electric hybrids

Non-plug-in hybrid cars, such as the Toyota Prius and Honda Insight, look and drive just like normal cars yet under the bonnet they contain a battery-powered electric motor as well as a regular combustion engine. All the energy is provided by petrol, but the electric part of the system enables the car to achieve significantly higher fuel efficiency.

The car charges its own battery when the brakes are applied – converting otherwise wasted kinetic energy into electrical energy – and also when the petrol engine is powering the vehicle along at high speeds. The battery's energy is then automatically used when lower speeds are required. The result is that hybrid cars can achieve over 60 miles per gallon – and exceptionally low levels of local air pollutants.

There are two main down sides with this kind of hybrid. First, they're ultimately powered by petrol, so their emissions remain higher than those of a car that can be plugged in and charged up via the mains. Second, they're currently quite expensive. If buying new, you can expect to pay around 10–20% more than you would for an equivalent non-hybrid model. For less money, you could get a super-efficient diesel – such as the Toyota Yaris or the Volvo S40 DRIVe – with roughly similar emissions. If you need a larger car, though, a hybrid may well be the best choice.

As some commentators have pointed out, the extra components in a hybrid car will add to the carbon footprint of its manufacture. This is likely to be true (it's hard to say for sure as good data is not readily available), but most experts agree that these extra emissions will be comfortably outweighed by the reduced fuel consumption. One much-publicized report claimed that the Toyota Prius had a bigger carbon footprint than an SUV if you took its manufacturing emissions into account. This conclusion, still sometimes repeated as if it were fact, seems to have been the result of a flawed calculation: the entire emissions involved in creating a new Prius factory were allocated to the small number of cars it had produced so far.

The Honda Insight's hybrid petrol–electric engine makes it one of the greenest non-plug-in cars on the market

Plug-in hybrids

Though not widely available at the time of writing, plug-in hybrid electric vehicles (PHEVs) are seen by many experts as the cars of the future. The idea is that you get all the convenience of a regular hybrid (you can fuel it with petrol when necessary) as well as the environmental and economic benefits of an electric car charged up by mains electricity.

The most promising PHEVs are those which have only a small petrol engine that exists solely to recharge the battery rather than actually power the car directly. In this arrangement, the engine is basically just an on-board generator that enables on-the-go charging.

Plug-in hybrids should start to appear on US forecourts in 2011 with the launch of the Chevrolet Volt, and in the UK in 2012 with the British version of the same car, the Vauxhall Ampera. Plug-in hybrid versions of the VW Golf and various other cars should also appear in 2012 or shortly thereafter.

Fully electric cars

Recharged via mains electricity – either at home or at a charging station when out and about – fully electric cars are the greenest automobiles available. Like most battery-powered devices, they have literally no harmful exhaust emissions, and, if charged up with electricity from renewable sources, their use creates virtually no carbon dioxide. Even if charged up with electricity created in fossil-fuel plants they're still significantly more eco-friendly than either petrol or diesel cars due to the higher efficiency of their motors and the regenerative braking they can employ.

Unfortunately, there are a few catches. First, you need a parking space near a plug socket: not ideal if your car lives on the street. Second, most of the electric cars currently available have limited speed and a battery capacity sufficient only for fairly short journeys. There are exceptions, such as the widely publicized Tesla and Lightning sports cars, but these performance cars, however impressive, are not designed for the mass market.

Tesla www.teslamotors.com
Lightning GT www.lightningcarcompany.co.uk

At the time of writing, the only fully electric cars available to buy in the UK are the G-Wiz (pictured) and the NICE Mega City. Both are small – especially the tiny G-Wiz, which is effectively a two-seater – and somewhat lacking in range and speed. The G-Wiz, which costs around £7000, can

Electric cars as energy storage plants

As we've seen, plug-in electric cars are typically more environmentally friendly than the petrol or diesel alternatives, even if the power they consume was generated by fossil fuels such as gas or coal. But they also offer another benefit: the capacity, in the long term, to help increase the proportion of renewable electricity installed on the grid. One problem with wind, solar, wave and tidal energy is that they're intermittent. This is an issue because in an electricity grid the supply of power at any one time has to be perfectly matched to demand (see p.76). One solution is electrical storage: finding ways to store power when too much is being generated so that it can be used in the future when the wind isn't blowing and the sun isn't shining. The more reliant on renewable energy we are, the greater the amount of electricity storage required.

Large-scale energy storage plants include split-level reservoir systems where water is pumped upwards when there's too much power being generated, and released back to the lower level, via hydroelectric turbines, when there's not enough. This approach works well, though there are very few sites suitable for split-level reservoirs.

Many experts now look to electric cars as a key energy storage device for the future. If half the cars on the road were electric, then one would expect millions of cars to be parked and plugged in at any one time – especially at times of peak electricity demand, such as the evening, when most people are at home. If the wind suddenly stopped blowing, these cars could all feed small amounts of electricity back into the grid from their recharged batteries, closing the gap between supply and demand. The car owners would be paid by the electricity companies for this service, helping to reduce the cost of owning an electric car.

The idea of using cars as electricity storage plants is part of a broad rethink of electricity demand and supply known as the "smart grid" or "energy internet".

do up to 40mph and can manage roughly forty miles before requiring a recharge (which takes a few hours). The Mega City, which starts at £12,000, has a similar top speed but a sixty mile range, more space and a better look. Neither would be suitable for long-distance drivers, but for city dwellers looking for an urban runaround, they're both practical and extremely inexpensive to run.

G-Wiz www.goingreen.co.uk
NICE Mega City www.nicecarcompany.co.uk

As with plug-in hybrids, fully electric cars with ranges and top speeds suitable for mass-market take-up should be sold from 2010 or 2011

onwards. The most promising models unveiled to date include the TH!NK City, hailing from Norway and capable of 60mph and 112 miles, and the all-electric version of the much-loved Mini, which is currently being made available in a trial rental scheme in the US. The "Mini E" (pictured) has a range of 150 miles and a top speed of 95mph.

Hydrogen fuel-cell cars

Instead of being charged up, fuel-cell vehicles generate electrical energy onboard via a catalytic process, usually the combination of oxygen (from the air) with hydrogen. The hydrogen can be made on demand from petrol or methanol, or, more commonly, generated elsewhere using electricity and stored on board in replaceable canisters.

Like other electric cars, fuel-cell vehicles help reduce air-borne pollution in towns and cities (water vapour is the only tailpipe emission) and they're extremely efficient in terms of CO_2 per mile. Trials using taxis and buses have already successfully demonstrated that the fuel cell can work well, and commercial models – such as the Necar, designed by DaimlerChrysler, Ford and Ballard – could conceivably be available to consumers within the next few years.

The main problem with fuel cells is that it takes a lot of energy (most of which currently comes from fossil fuels) to make the hydrogen in the first place. Another problem is that the fuel cells themselves are fairly large, and therefore hard to accommodate in small cars. In addition, a whole new hydrogen infrastructure – production plants, distribution channels and refilling stations – would need to be rolled out to make these cars function on a mass scale. For all these reasons, most experts today back all-electric vehicles and plug-in hybrids as the cars of the future, rather than the hydrogen fuel-cell vehicles that seemed to hold a lot of promise a few years ago.

Motorbikes and mopeds

As the diagram on p.145 shows, motorbikes vary widely when it comes to CO_2 emissions. A modern moped with two passengers can be as climate-friendly as some trains. At the other end of the spectrum, a 1000cc powerbike may consume as much fuel per mile as a typical family car.

If you live in a city, however, it's worth knowing that many motorbikes have comparatively bad emissions of poisonous gases and particulates. One Swiss study found that even a Vespa scooter from 1997 was far worse than a typical modern car in terms of poisonous gas outputs. More recent bikes tend to have lower harmful emissions, but many are still worse than cars in this respect.

Electric motorbikes and mopeds

As with cars, the greenest motorcycles are electrically powered. Electric mopeds are already widely available, with decent models starting at around £1500 – a similar price to a new brand-name petrol moped. In terms of emissions and cost, driving an "e-scooter" for thirty miles is equivalent to leaving a 100W light bulb on for just a few hours. And, of course, such vehicles produce no poisonous exhaust fumes. Some models – such as the EVT168 (pictured) – even look flashy, too. The down side is that most current e-scooters can only manage about 30mph (or even less when driving up a hill or carrying a pillion passenger) and they need to be recharged every thirty miles or so. For more information, visit:

Electric Bikes www.electric-bikes.com

In the performance bike sector, electric models are starting to appear that match even the best petrol motorcycles for power and performance, and offer a much longer range. One example is the Mission One, which can do up to 150 miles per hour (with maximum torque available at any speed) and drive for 150 miles on a single charge.

Mission One www.ridemission.com

Buying fuel
Petrol, diesel, biofuels and LPG

Even if you don't upgrade to a greener vehicle, you may want to consider what you put in your existing one. That might mean using a more eco-logically friendly alternative to petrol or diesel, such as locally produced biofuel or LPG. Or it might simply mean favouring certain petrol brands over others.

Biofuels

Biofuels are plant-based alternatives to petrol or diesel. The idea is that they're more climate-friendly than conventional fuels because the CO_2 released when they're burned is no greater than the CO_2 they removed from the atmosphere as the plants grow. In truth, though, the biofuels currently available often have a larger carbon footprint than petrol or diesel. In addition, since these first-generation biofuels are made from food crops such as corn and oil-seed, they put pressure on food supplies and incentivize the clearing of rainforest to make way for agricultural land. There are two main kinds of biofuel: biodiesel for diesel engines, and ethanol for petrol engines.

Biodiesel

The term diesel, by definition, simply means a fuel for powering a diesel engine. Today it generally refers to a type of petroleum, but when Rudolph Diesel invented his super-efficient combustion engine at the end of the nineteenth century, he envisaged the fuel of the future coming from plants; he famously ran one prototype on peanut oil. Today, with concerns about climate change mounting, so-called biodiesel is making a come-back. The fuel is very similar to vegetable oil and is produced from the same kinds of plants – oil-rich sources such as sunflower, palm, rapeseed and groundnut. It can also be made from animal fats or vegetable oils recycled from takeaways and other restaurants.

Whether biodiesel is an environmental blessing or curse depends on its source. Fuel made from recycled vegetable oil is unquestionably green: it requires minimal processing and the oil would otherwise be basically worthless. So if this kind of biodiesel is available in your local area, by all means use it. Before filling up, check with the manufacturer of your car whether they approve the use of such fuel, either neat or in a mix, or you could invalidate your engine warranty.

Unfortunately, there will only ever be enough recycled frying oil to power a tiny proportion of cars and trucks; the rest of the world's biodiesel comes from food crops. This is where the environmental and ethical problems kick in. Diverting any crop from food production to fuel production helps drive up food prices, which is bad news for many of the world's poorest people. Moreover, some oil crops such as palm are well suited to being grown in rainforest climates, and large areas of ecologically precious and carbon-rich forest in Southeast Asia have already been cleared to make way for palm oil plantations.

Vegetable oil as a fuel

"Proper" biodiesel is one thing, but what about those stories in the press about people filling up with plain old vegetable oil in supermarket car parks – people such as Daniel Blackburn (pictured), who made the headlines in 2003 by using veg oil to motor all the way from John O'Groats to Land's End?

Daniel Blackburn

It's true that, after a conversion costing around £500–1000, many diesel engines will run perfectly well on standard cooking oil. As long as you declare what you're doing and pay the relevant tax, it's perfectly legal and fairly economical (see vegoilmotoring. com for details).

But is it green to use veg oil as a fuel? On the one hand, it's probably better than regular biodiesel as it is said to require less energy-intensive processing. In addition, it's possible to choose non-tropical oil types such as sunflower and thereby avoid ecologically problematic palm oil. Ultimately, though, if you buy vegetable oil for your car, you'll still be increasing the overall demand for oil crops and taking food out of the system, so the benefits are far from clear cut.

Bioethanol

Ethanol, more commonly known as alcohol, is a substitute for gasoline. All petrol cars can handle a small amount of ethanol mixed in with their regular fuel (most run well with a 10% mix) and some "flexifuel" vehicles can handle any mix up to and including neat ethanol. When the ethanol is produced from plants, it's often referred to as bioethanol. This biofuel hasn't caught on in a huge way in the UK, though it's big business in Brazil, where it's made from sugar, and the US, where it's most commonly made from corn. American ethanol refineries now turn as much as 5% of the world's food grains into car fuel. This, according to research by the World Bank and others, was a significant contributor to the spikes in food prices that caused rioting and hunger in many developing countries in 2008.

The environmental impact of bioethanol depends on the crop it's produced from and how that crop was grown. Most studies agree that when you consider the entire farm-to-fuel footprint – including fertilizer production and the nitrous oxide emissions from the agricultural processes – then ethanol produced from corn causes more CO_2 emissions than petrol. As this conclusion becomes ever more widely accepted, advocates of ethanol fuel are increasingly pinning their hopes on second-generation bioethanol made from cellulose. This stringy molecule is found in agricultural wastes as well as wood, switchgrass and other plants that can be grown on non-agricultural land with no synthetic fertilizers. However, it's a scientific challenge to turn cellulose into ethanol without using too much energy or spending too much money, so it will probably be some time before this second-generation bioethanol is available in the UK.

Liquid petroleum gas

LPG (liquid petroleum gas) is mainly propane, as used in camping stoves and stand-alone gas heaters. A byproduct of oil refining and natural gas extraction, it's a fossil fuel but it has lower CO_2 emissions than petrol and also results in lower levels of poisonous fumes. Most petrol-powered cars can be converted to run either solely on LPG or on both LPG and petrol. The very best emissions savings are made by LPG-only conversions, though a dual-fuel system is more convenient given that LPG isn't available at every filling station. You can check availability in your area at DriveLPG or order a paper LPG filling-station map from Go Autogas:

DriveLPG www.drivelpg.co.uk
Go Autogas www.go-autogas.com

Smart cars, such as this cabriolet, are a popular choice for LPG conversion. The result is a very low-emission car that's inexpensive to run.

Running a car on LPG results in CO_2 savings of around 15–20%. In other words, an LPG car is about as climate-friendly as a diesel car. However, whereas diesel is a fairly dirty fuel in terms of local air pollutants, LPG is very clean, and therefore a more ideal choice for urban use.

Conversion to LPG usually costs £1500–2000, but once it's done you make big savings on fuel thanks to government subsidies. If you drive the car a typical distance each year – 9000 miles – then it would only take a few years to repay the initial investment and the savings over the life of the car might be in excess of £10,000. In addition, most cars powered by LPG are exempt from the London congestion charge and cost a minimal amount to tax. Since the fuel is so clean, you may also save money through better engine longevity.

In terms of performance, early LPG conversions sometimes caused cars to become less zippy, though this isn't usually a problem today. Similarly, while early conversions tended to require a gas tank in the boot, where it took up useful space, modern conversions often place the tanks underneath the car.

Which oil company is greenest?

In a world threatened by climate change, there isn't any such thing as environmentally friendly petrol or diesel. But if you are going to buy these fuels, is there perhaps a case for avoiding certain oil companies and favouring others, based on their respective record on environmental issues? Arguably, yes, though the differences between the various companies seem less stark today than they did a few years ago.

Back in the 1990s, BP (now part of BP Amoco), Shell (of Royal Dutch/ Shell) and Exxon (Esso in the UK) were all key players in the Global Climate Coalition, a lobby group formed in 1989 as the prospect of global diplomatic action on climate change appeared on the horizon. Along with lobbying at UN meetings, the coalition became an oft-quoted presence in US news reports and financed anti-Kyoto commercials warning that "Americans would pay the price" for the treaty. The GCC began to fracture with the departure of BP in 1997 and Shell in 1998. By 2001, it was history, though arguably it had served its purpose and was no longer necessary. A 2001 memo written to Exxon by the US undersecretary of state, Paula Dobriansky, and later obtained by Greenpeace, states that George Bush rejected Kyoto "partly based on input from you [the GCC]".

Even after the GCC ceased to exist, Exxon continued to give millions of dollars to think tanks and other organizations that sought to challenge the scientific and economic case for emissions cuts. These contributions seemed to pay off. In the autumn of 2003 (within days of World Health Organization scientists suggesting that more than 150,000 people already die each year from climate change impacts), leaked emails and documents showed that the Bush administration had sought the help of one Exxon-funded think tank – the Competitive Enterprise Institute – to try to undermine and dilute the predictions of its own government scientists. Exxon's donations to climate sceptics continued, and in 2006 the UK's Royal Society took the unprecedented step of writing an open letter to the company appealing for it to stop.

Today, Exxon acknowledges the existence of human-induced climate change but claims that policy responses need to be "paced" (unlike most climate experts, who advocate rapid and radical policy action). The company has put a token amount of money into renewables and climate research but its actions have so far been insufficient to persuade climate campaigners to call off their consumer boycott. Find out more via:

Exxon/Esso www.esso.co.uk
StopEsso www.exxposeexxon.com

Even while Exxon was brazenly funding climate change denial, some other big oil companies were spending millions trying to redefine themselves as trailblazers of corporate responsibility. BP very much led this trend, swapping its shield logo for a green and yellow flower and announcing that its initials now stood for "Beyond Petroleum". Shell followed close behind with numerous green and ethical initiatives, keen to bury associations with Ken Saro-Wiwa and other anti-Shell campaigners who were executed in Nigeria in the mid-1990s. Sadly, neither company has lived up to its environmental promises.

In 2008, Shell's reputation among environmentalists went from bad to worse when it pulled out of the London Array, the UK's biggest wind-farm project, and boosted its investment in the Canadian tar sands – a source of "unconventional" oil that has even higher lifecycle emissions than crude. Then, in 2009, only weeks after placing prominent advertisements about its commitment to a sustainable future, Shell announced that it was to stop building wind and solar installations around the world. BP's green credentials haven't collapsed quite so dramatically, though things have gone downhill since the departure of chief executive Lord Browne in 2007. Like Shell, BP has now invested in tar sand extraction and, despite ubiquitous marketing trumpeting the company's commitment to renewables, clean energy sources remain only a tiny proportion of BP business. At the time of writing, BP has just announced hundreds of redundancies from its Solar division.

All in all, then, though there may still be a case for shunning Esso, there certainly isn't a green halo to be had by favouring its competitors.

Driving less
Walking, cycling and car sharing

Efficient vehicles and lower-carbon fuels are all well and good, but probably the best way to reduce the environmental impact of our driving is simply to do less of it. The most effective ways to reduce mileage are to stop commuting by car and to make fewer long trips. Short hops to the local shops may seem more wasteful, since they can be made on foot or bicycle, but it's the longer journeys that make the most difference.

How green is cycling and walking?

It's often assumed that cycling or walking are zero-carbon forms of transport. But that isn't quite true. After all, taking extra exercise increases the amount of food we need, and in some cases the growing and distributing of that extra food causes a surprising amount of emissions. In 2007, environment writer Chris Goodall caused a stir by demonstrating that walking could be slightly more polluting than driving if the extra calories came from milk, or four times more polluting if they came from beef. Goodall's argument forcefully highlights quite how carbon intensive some of our foods are, but the idea that walking may be less green than driving doesn't add up. For one thing, each of us needs to take exercise to stay healthy. If we take that exercise by – among other things – making short trips on foot rather than by car, then we'll be displacing fossil fuel use. If we drive everywhere, and perhaps get our exercise in the gym, we'll be increasing fossil fuel use.

What about cycling? According to physicist David MacKay, riding a bike requires on average about half as many calories as walking, making it the most energy efficient of any form of transport, with the possible exception of a packed electric train.

Lift sharing and car clubs

If you want to get shot of your car but don't want to rely solely on public transport, consider looking into car sharing and car clubs. This is not only environmentally sound, but may also make sense financially.

Either for a regular commute or a one-off drive, car sharing is based on the simple rationale that one car carrying, say, three people is three times less polluting, congesting and expensive than three cars carrying one person each. Although the UK has been much slower to grasp this fact than much of continental Europe, the car-sharing movement is slowly taking off. To search for a ride, or to find passengers to share the cost of fuel, see:

Freewheelers www.freewheelers.co.uk
Liftshare www.liftshare.com
National Car Share www.nationalcarshare.co.uk

Car clubs or car pools are something else entirely. You don't actually own a car but have access to a communal one situated within a few minutes' walk of your house. Typical fees are £4–6 per hour, or £50 for 24 hours. Popular clubs in major cities include:

City Car Club www.citycarclub.co.uk
Street Car www.streetcar.co.uk

To find your nearest car clubs, see:

Car Plus www.carclubs.org.uk

Air travel & alternatives

Exactly how bad is flying?

There's no way around the fact that flights are bad news for the environment. It's not just that planes are worse than most other forms of transport in terms of the greenhouse impact per passenger mile. Just as important is the simple fact that flying allows us to travel a far greater number of miles than we otherwise could or would do. Thanks to these two factors, individual trips by air can have a remarkably large carbon footprint. Two seats on a return trip from London to San Francisco produces the equivalent impact of around five tonnes of CO_2. That's comparable to ten months of gas and electricity use in the typical UK home, or to driving 20,000 miles in an average car.

The total impact of flying

As the aviation industry is always keen to point out, planes account for only around 1.5–2% of global CO_2 emissions. However, this figure is somewhat misleading. For one thing, most flights are taken by the wealthy, so in developed countries the slice of CO_2 emissions caused by flying is higher – around 6.3% in the UK, according to Department for Transport figures for the year 2005. Even this figure underplays aviation's environmental footprint, and not just because the number of flights has risen since 2005. There are at least three other reasons why 6.3% is likely to be a strong underestimate.

First, the total global warming impact of each flight is thought to be around twice as high as the CO_2 emissions alone, for reasons discussed later in this chapter. Second, the figures are slightly skewed in favour of

British travellers. The standard way to account for the emissions for an international flight is to allocate half to the country of departure and half to the country of arrival. But UK residents take up two-thirds of the seats on the average plane landing at or taking off from a British airport. This means the official statistics are effectively offloading the emissions of British holidaymakers and businesspeople onto the countries they're visiting. Third, the aviation industry causes emissions over and above those of the planes themselves. The processing and transportation of the aviation fuel, and the manufacture and maintenance of planes, airports and support vehicles all create extra carbon dioxide.

There's not enough data to say for sure, but it seems likely that aviation's true impact in the UK is around 13–15% of total greenhouse gas emissions. If that still sounds fairly low, compared to the massive amounts of attention heaped on aviation by climate change campaigners, bear in mind that most people in the UK don't regularly fly. The average British resident takes a short-haul flight only every two years, and a long-haul flight only every five years. In other words, the air travel of a minority of regular flyers causes a substantial slice of UK emissions.

Looking forward, it's very hard to reconcile the British government's plans for increased aviation capacity with its plans for carbon cuts. The UK is seeking to reduce its emissions by 80% by 2050, relative to 1990 levels. At the same time it predicts a rise in the number of flights sufficient to use up more than half of the remaining 20% of emissions. This fact, combined with a lack of promising low-carbon plane technology, helps explain why air travel has become such a heated topic in the climate change debate.

Beyond CO_2

The impact of planes on the climate is complicated and not perfectly understood. The CO_2 emissions are straightforward enough, but plane engines also generate a host of other outputs, including nitrous oxide, water vapour and soot. At flying altitudes in the upper troposphere and lower stratosphere, these outputs produce a range of climatic effects. For example, the nitrous oxide causes the formation of ozone – a greenhouse gas that warms the local climate – but at the same time undergoes reactions which destroy methane, thereby removing another greenhouse gas from the atmosphere.

Even more complicated is the impact of soot and water vapour, which together can cause contrails (vapour trails) and in cold air can lead to the

Aircraft vapour trails over the southeast US
NASA Langley Research Centre

formation of cirrus clouds. The science surrounding this topic is not yet rock solid, but researchers believe that contrails add to the greenhouse effect – especially at night, when their tendency to stop heat escaping from the Earth isn't offset by their tendency to reflect incoming sunlight.

Today, most experts favour a CO_2 "multiplier" of around two. In other words, they believe that the total impact of a plane is approximately twice as high as its CO_2 emissions. The exact multiplier will always depend on the individual plane, the local climate and the time of day.

What about greener planes?

A number of technologies designed to reduce the environmental impact of flying have been researched, tested and implemented. However, compared to greener cars, where the technologies are proved and the carbon saving huge, the potential for eco-friendly flying looks very limited. There will be some further gains in engine efficiency over the coming decades, and larger planes with more seats will allow slightly lower emissions per passenger. But there is nothing in the pipeline with the transformative potential of the electric car. The problem is that electric motors can't produce enough power to get a plane off the ground, so the only alternative to regular kerosene-based aviation fuels are special kinds of biofuels. These aren't an ideal solution. As the previous chapter explained, biofuels can be environmentally problematic in themselves, and anyhow it would take a huge chunk of the world's arable land to grow enough crops to fuel all the world's planes. (A back-of-the-envelope calculation suggests it might require as much as a fifth of all cropland.) Even if it were possible or desirable to devote this much land to fuelling planes, it wouldn't do much to get around the problem of contrails.

With all this in mind, the only likely way in which air travel's climate impact could be substantially reduced would be for governments to reduce passenger numbers – either by somehow rationing flights or by taxing them sufficiently to discourage regular flying.

Boeing and bombing

As if the environmental impact of flying wasn't enough, there is the broader ethical concern that most passenger aircraft are produced by arms manufacturers. A growing number of consumers are opting for ethical banks, very often specifically because they want to be sure that their savings aren't invested in arms companies. But, for anyone who travels by air, it's not so easy to completely separate your wallet from the budget sheets of "defence" firms. Next time you fly, have a look to see who produced your airborne home for the next few hours. With very few exceptions, it will be either Boeing or Airbus.

Boeing is one of the world's largest arms companies, whose annual turnover of around $50 billion is in no small part generated from selling military equipment to all kinds of governments, including those with very poor human rights records. According to an investigative report in *Mother Jones* magazine, recent Boeing sales include warplanes to Indonesia, Israel, Kuwait and Saudi Arabia; attack helicopters to Egypt; and missiles to Turkey. Boeing is even "unethical" according to the low moral codes of the arms industry: in late 2003 its chairman, Phil Condit, resigned after, as the Associated Press put it, "months of ethical controversies over the aggressive methods it used to obtain lucrative defence contracts".

Airbus, meanwhile, is owned by British Aerospace and other major European arms manufacturers. All in all, the tie between commercial aircraft and military equipment is so entrenched that – as Noam Chomsky has written – many passenger planes are essentially modified bombers. There's not much that consumers can do about this link, but if you feel particularly strongly about the arms industry it may tip the balance and make you decide to choose another form of transport whenever possible.

What individuals can do

For anyone concerned about global warming, cutting back on air travel is an obvious goal. This might mean giving up flying altogether or it might mean taking fewer flights and picking destinations that are closer to home. It's true that short flights tend to be more harmful to the climate per mile travelled than long-haul flights are (because they have more empty seats, and because taking off and landing burns more fuel than cruising) but this doesn't change the fact that the further you travel, the greater the emissions that will result.

If you do fly, you can in theory make some small difference to the carbon impact by favouring day-time flights. This at least ensures that any contrails caused by the plane will reflect some sunlight away from the Earth in addition to locking warmth into the atmosphere. Also consider limiting your luggage. It doesn't make much sense to worry about air-freighted vegetables one minute if the next minute you're lugging a huge suitcase onto an airport check-in desk. Each extra kilo that a plane carries marginally increases its fuel burn.

Finally, you might want to consider which airlines you use. People often assume that budget flights are somehow more eco-unfriendly than expensive ones. In fact, the opposite tends to be true. Budget airlines pack more passengers on each flight and typically have younger, more fuel-efficient fleets than longer-established airlines. Indeed, the least eco-friendly tickets of all aren't the cheapest but the most expensive. Business-class and first-class seats take up more space on the plane, thereby reducing the number of people who can fit on each flight. In addition, by virtue of being disproportionately more expensive, high-end tickets effectively subsidize inefficient airlines, helping them stay afloat.

As for travel agents, they're all much of a muchness, with the one exception of North South Travel. This friendly, efficient service donates all its profits to charitable projects in Africa, Asia and Latin America. That won't change the emissions of your flight, but it may at least bring some positive benefits in other ways.

North South Travel www.northsouthtravel.co.uk

Ultimately, though, there's no way around the fact that even a couple of flights a year can make it almost impossible to get your overall carbon footprint down to a sustainable level. So it makes real sense to limit your air travel and to opt for alternative modes of transport where possible.

Greener to go by train?

As a rule, taking the train instead of the plane will substantially reduce your carbon emissions – perhaps by a factor of five to ten on a domestic trip. The benefits will be somewhat reduced as the journey gets longer. That's partly because, as we've seen, shorter flights are more polluting per passenger mile than longer ones, but it's also because long train journeys usually necessitate sleeping onboard. Sleeper cars usually carry fewer passengers than regular carriages, so their emissions per passenger are higher. If, as is common in some countries, the train is powered by diesel rather than electricity, then the emissions will be higher still. Indeed, a diesel sleeper train travelling a long distance might emit nearly as much CO_2 per passenger as a plane. Even then, the train will typically be greener once you consider the plane's non-CO_2 warming effects, but the fact remains that long-haul rail is not as inherently eco-friendly as some press articles would have us believe. A trip on the Trans-Siberian Railway may make for a great adventure, but it's likely to have a far larger carbon footprint than a holiday involving a short-haul flight.

Of course, in the case of electric trains, the emissions depend on the fuels used to generate the power. For this reason, train travel in France, with its largely nuclear power grid, is even more climate friendly than train travel in most other countries. This explains why Eurostar can legitimately claim that a London to Paris train ride causes less than a tenth of the CO_2 emissions of the same trip made by air. Indeed, as a general rule, travelling by rail from the UK to France – or to Italy and Spain via France – offers particularly large carbon savings.

Unfortunately, almost every long-distance train journey will cost you far more than flying would. Indeed, the difference in price is often so great that for some unavoidable trips it would arguably make sense to take the plane and spend the savings on something more environmentally beneficial than a train ticket. For example, while two people might be able to fly from London to Rome and back for £100, the same trip by train might cost £700. In this example, the extra £600 would be saving less than a single tonne of CO_2. At this price, reducing emissions by train travel is almost a hundred times more expensive than reducing emissions by buying offsets. Even if you don't believe in offsetting, the £600 would save incomparably more carbon if spent on home insulation, say, or on buying and destroying dozens of tonnes' worth of EU carbon credits.

On the other hand, flying is an inherently high-carbon activity that for many people embodies our inability to take climate change seriously.

For this reason, some green-minded individuals refuse ever to fly. If the alternatives are too expensive, it's always possible to holiday closer to home, after all.

To find times and costs for train and boat trips from the UK to almost anywhere in the world, visit Seat61, a remarkably definitive website run by railwayman and international train traveller Mark Smith:

The Man in Seat 61 www.seat61.com

Travelling long-haul by boat

Most short-haul trips can be fairly easily managed by train, coach or car. But some journeys, including any trip from Europe to the US, can be made only by air or sea. There are two options for long-haul boat travel: cruise ships and freight boats. Unfortunately, neither offers a solution that's both practical and green.

Europe to the US by ocean liner

The cruise liner that most frequently crosses the Atlantic is the *Queen Mary 2*, which runs regularly from Southampton to New York, with a journey time of around six days each way. Sadly, this unquestionably glamorous service is not in the least bit eco-friendly. Basic physics dictates that moving a large ship through water at cruise-ship speeds takes a huge amount of energy. In the case of the *Queen Mary 2*, this energy is provided by approximately three tonnes of heavy oil, plus up to six tonnes of marine gas oil, each hour. In other words, it takes 400–1200 tonnes of fuel to get the ship from the UK to the US. All that fuel propels just 2600 passengers, making the overall carbon footprint of each ticket substantially more carbon intensive than a seat on a plane.

Europe to the US by freight ship

A second way to get across the Atlantic without flying – or indeed to reach other faraway destinations – is to buy passage on a freight ship. Many container ships offer a few fairly comfortable berths at any one time (generally enough for twelve passengers). Although these ships burn plenty of fuel, they're not as energy-profligate as cruise ships. More importantly, they're working boats that will be making the same trips regardless of whether a few passengers come along for the ride. But container-boat travel isn't for everyone. One problem is the very limited number of routes. To get from London to New York, for example, you'd probably catch the Antwerp–Liverpool–Philadelphia boat and travel by train at each end. Second, the slow speeds make for very long travel times. A trip from the UK to the US might take a couple of weeks: a long period to spend on a ship full of metal crates and with little in the way of entertainment. Despite these disadvantages, cargo-boat travel isn't cheap. Once you factor in the trains at each end, you might spend almost as much getting from London to New York on a freight ship as you would catching the *QM2*. If none of that puts you off, you can find more information here:

Freighter Cruises www.cruisepeople.co.uk

Ecotourism
Do green long-haul holidays exist?

The most obvious green consideration when choosing a holiday is the greenhouse emissions of the plane, car or train that will transport you to your destination. As we've seen, even a short-haul flight has a fairly large carbon footprint. However, our holidays also have an environmental (and indeed social) impact in the countries we visit. In many cases, it's fairly clear that the impact is a negative one. From coastal Spain to Goa, many resort areas have had their biodiversity decimated by large-scale tourist development. And in countries that lack decent waste disposal, tourist junk – from sun-lotion bottles and food wrappers to toilet paper – may end up in rivers that both people and wildlife depend on.

Nonetheless, some holiday companies sell trips specifically on the basis of a green or ethical angle. This sector of the travel industry – generally operating under the banner of "ecotourism" or "responsible travel" – has grown from a niche market into a major sector, embraced both by tourism industry bodies and by the UN (who named 2002 the International Year of Ecotourism, complete with a World Ecotourism Summit in Quebec).

So what exactly *is* ecotourism? According to the International Ecotourism Society, the term refers to "responsible travel to natural areas that conserves the environment and sustains the well-being of local people". Or, as the World Conservation Union put it, ecotourism describes "environmentally responsible travel … to relatively undisturbed natural areas … that promotes conservation, has low negative visitor impact

[and] provides for beneficially active socioeconomic involvement of local populations."

Advocates of ecotourism claim that certain types of trips can contribute a great deal both to conservation and to the economic empowerment of people in remote regions. Certainly it's true that nature travel can provide a perfect incentive for countries to look after their environments. If tourists are coming to see beautiful forests or wildlife, these very elements become valuable assets worth protecting. And in countries where few other employment opportunities exist, the travel sector can create jobs that offer an alternative to ecologically damaging work such as tree-felling or small-scale mining.

Similarly, it's true that tourism can provide a valuable source of foreign income in developing countries. Indeed, the very poorest countries – including many of those which are popular for ecotourism – sometimes stand to benefit the most. A 2001 report from the United Nations Conference on Trade and Development pointed out that "International tourism is one of the few economic sectors through which LDCs [Least Developed Countries] have managed to increase their participation in the global economy. It can be an engine of employment creation, poverty eradication, ensuring gender equality, and protection of the natural and cultural heritage."

Whether an ecotourism trip can ever help protect an endangered ecosystem sufficiently to outweigh the carbon emissions involved in flying to get there is an open question. If a small number of tourists per year visit and thereby encourage the protection of a large swathe of rainforest, and they're shown around by a guide who is knowledgeable of and sensitive to local needs and concerns, then it may well be that the net impact will be positive. If, on the other hand, a hundred thousand people fly on "eco" trips each year to visit Australia's Great Barrier Reef – which is endangered by global warming rather than a lack of conservation rules – then it's hard to see how the benefits could outweigh the cons.

Of course, what starts as a trickle of travellers can often end up as a flood. So even with small-scale ecotourism there's a risk that a few adventurous and well-meaning travellers could effectively be opening up the world's most fragile environments to unsustainable levels of tourism. According to a recent report by Conservation International and the environmental wing of the UN, in the 1990s alone, leisure travel to the world's "biodiversity hotspots" (those areas with richly diverse but delicate ecologies) more than doubled, with rises of more than 300% in Brazil,

Nicaragua and El Salvador, 500% in South Africa and 2000% in Laos and Cambodia.

When visitor numbers rise, ecotourism needs to be very carefully managed, even if it is succeeding in encouraging conservation. In the case of safari holidays, for example, the tourists and their guides may sometimes cause harm to the very animals they have come to see. As Philip Seddon of New Zealand's University of Otago in Dunedin recently told *New Scientist*: "Transmission of disease to wildlife, or subtle changes to wildlife health through disturbance of daily routines or increased stress levels, while not apparent to a casual observer, may translate to lowered survival and breeding."

Unfortunately, it's hard to know in advance which travel operators are sensitive to these kinds of concerns. Ecotourism has no legally binding definition, so there's nothing to stop an unscrupulous travel agent from slapping the label on any nature-focused holiday, regardless of the damage it may cause. As one travel website puts it, "An eco-lodge may dump untreated sewage in a river, and still call itself eco simply because it is located in a natural setting."

All told, probably the best advice if you do plan to embark on any ecotourism is to favour small-scale, long-standing operators whom you can quiz in advance about their approach. There are also specialist travel agents which operate screening criteria for the companies that they work with – the best known being responsibletravel.com. But even if you find a perfect operator, be aware of the inherent contradiction of attaching the word "eco" to any holiday involving an environmentally destructive long-haul flight.

Part IV

Food

Food: the basics
Meat, dairy and eggs
Fruit and veg
Where to buy food

Food: the basics

Climate change, organics, food miles, GM and fair trade

13

No consumer area is so ethically charged as food, which is at the centre of debates ranging from rainforest loss, soil degradation and nitrous oxide emissions through to public health, biotechnology and farmer exploitation. This chapter covers some general food issues, including organic production, food transport and fair trade. The following chapters focus in on particular types of foods and the various places where we can buy them. First, though, let's quickly take stock of food's contribution to climate change, and explain in brief how to choose a low-carbon diet.

Food and climate change in brief
The footprint of food and how to help reduce it

No one knows exactly what slice of the world's total carbon footprint is attributable to food, but the figure is probably at least ten percent and maybe as much as a quarter. The reason there's so much uncertainty is that the footprint of food is formidably complex. It's easy enough to roughly measure the oil and electricity consumed by farms, vans and shops, and to tally up the huge volumes of natural gas used to produce synthetic fertilizers. It's harder to precisely understand the potent greenhouse gas emissions created through biological processes. These extra emissions

– which include nitrous oxide generated by fertilized soils and animal manure, and methane escaping from rice paddies and the digestive tracts of cows and sheep – are hugely significant. According to some estimates, nitrous oxide alone accounts for the majority of the climate impact of British farms.

To make things more confusing still, food production is intricately bound up with another pivotal environmental issue: the destruction of tropical rainforests. Deforestation causes more CO_2 emissions than every vehicle in the world combined, and the main cause of the destruction is the clearing of land to grow crops and graze cattle. Even if we don't personally consume crops grown on cleared patches of rainforest, our food choices still affect the global supply and demand for land, so they can't be entirely separated from this destruction. As we'll see in the following chapter, one UN body has estimated that, if the impact on deforestation is taken into account, livestock and their feed alone cause almost a fifth of human-created greenhouse gases.

What a low-carbon diet looks like

The following three chapters give lots of advice on reducing the carbon emissions and other negative environmental impacts of your food purchases. For readers who just want the key tips for cutting their carbon footprint, however, here's a brief summary of how to achieve a low-carbon diet, with pointers to the pages in this book where you can find out more on each topic. The following aims are arranged roughly in order of priority.

▶ **Minimize animal products** Animal products generally have high footprints (see p.209). Beef, lamb and cheese are particularly carbon-intensive.

▶ **Minimize waste** Buying only the food you need (see p.116) reduces waste and demand for land, while composting food scraps avoids them turning into methane (see p.132).

▶ **Favour local, seasonal fruit and vegetables** Non-exotic fresh produce, grown locally and in season are typically the lowest-carbon fruit and vegetables available (see p.239).

▶ **Avoid air-freighted foods** Premium vegetables transported by plane have incomparably high footprints compared to those imported by ship, though boycotting them raises an ethical conundrum (see p.196).

▶ **Shop near home or not too often** There's little point choosing low-carbon foods if you need to drive miles to buy them (see p.196). Shopping locally or in bulk can help reduce car use.

▶ **Cook from scratch** Buying and cooking ingredients will typically use less energy overall than choosing a readymeal that will have been chilled or frozen since leaving the factory.

▶ **Favour organic foods** In general, organic foods have smaller carbon footprints than the conventionally grown alternatives (see p.188) but the point is debated and there are certainly exceptions.

▶ **Use a gas cooker when possible** As chapter six explained, gas cookers typically cause far fewer carbon emissions than electric cookers or microwaves (see p.72).

▶ **Cook quickly** Obviously, the longer a cooker is running (and the higher the temperature), the more energy it consumes. Hence the very greenest meals are raw ones, such as salads and sandwiches.

▶ **Avoid heavy packaging** Plastic packaging is unattractive but its carbon footprint is typically lower than heavier materials such as glass or paper (see p.258).

Organic food and drink
Is it greener and more ethical?

Agriculture in the UK – and indeed the wider world – has undergone a remarkable transition in the last sixty years. In the aftermath of World War II, Britain's drive to produce as much as possible as quickly as possible saw massive state subsidies awarded to the biggest, most industrialized farms. Small producers that relied heavily on manpower and produced a range of crops and animal products were gradually eclipsed by larger businesses that focused on specific crops or meats and relied heavily on technology, economies of scale and agrochemicals (the collective name for fertilizers, pesticides and other synthetic inputs used in farming).

One impact of this process of industrialization has been a massive decrease in the number of farm workers. Indeed, there are now nearly half a million fewer agricultural jobs in the UK than there were a few decades ago – and although the amount of food we import has played a part in this, the main factor has been the increased efficiency allowed by technology

Soil

Soil may not be the most glamorous substance in the world, but it's surprisingly fascinating stuff and it lies at the centre of debates about organic and industrial farming. Despite its dull appearance, earth is absolutely packed full of life. The UN's Global Biodiversity Assessment suggested that a single gram of soil "could contain 10,000 million individual cells comprising 4000–5000 bacterial types, of which less than 10% have been isolated and are known to science", and that's before you consider snails, earthworms, termites, mites and other invertebrates. But soil not only *contains* life, of course, it also gives it: without fertile earth, there'd be no plants, no animals and no humans.

The world is covered by an extremely thin and delicate skin of this fertile dirt. (If the world was the size of a football, the layer of soil would be many thousands of times thinner than a piece of paper wrapped around the ball.) And yet humans have been rather cavalier in their treatment of this precious resource, battering it with intensive, chemical-heavy agriculture, urbanization and deforestation. According to a report by the UK's Royal Commission on Environmental Pollution, around 10% of the world's total soil has been lost through human-induced causes. Perhaps even more remarkably, in just the last fifty years, nearly a third of the world's arable crop land has been abandoned due to soil erosion. Considering that it takes around 500 years to form just a few centimetres of soil in agricultural conditions, we're clearly losing our life-giving earth far more quickly than it can be replaced, and in the process endangering the future world's ability to feed itself.

It was precisely these kinds of worries that led to the establishment of the organic movement. Hence the fact that its longest-standing organization is known as the Soil Association.

Corbis

Source: All figures quoted in Biodiversity for Food and Agriculture, *a report by the Sustainable Development Department of the United Nations Food & Agriculture Organization*

and chemicals. Depending on who you ask, this reduction of the farm workforce is either a tragic loss of tradition or a good thing, since people in developed countries no longer want or need those kinds of jobs.

Another impact of subsidized industrial agriculture has been a fall in the price of the food in our shops. The average UK household is now estimated to spend less than one tenth of its total budget on food. This compares with around one third just fifty years ago, and that's despite the fact that we're eating more expensive and exotic ingredients than ever before. Of course, some of the relative fall in our food spending is down to increasing wages and the fact that food prices are kept artificially low by tax-funded farm subsidies. But even taking these factors into account, food has become significantly less expensive as farming has got more industrial.

Most people would agree that cheaper food isn't a bad thing, the rise in obesity notwithstanding. But according to many commentators, we're not getting something for nothing. The hidden costs of industrial farming – to the environment, to human health, and long-term food security – are big and getting bigger. In the words of the staff writers at science journal *Nature*, "Mainstream agronomists now acknowledge ... that intensive farming reduces biodiversity, encourages irreversible soil erosion and generates run-off that is awash with harmful chemicals – including nitrates from fertilizers that can devastate aquatic ecosystems." And that's before you consider carbon emissions, animal welfare and the inherent wastefulness of some intensive meat farming.

Some saw these problems coming right from the beginning. As agricultural industrialization took off, a small bunch of philosophically minded farmers, worried primarily about the potentially damaging effect on soil, planted the seeds of the organic movement, which, after decades on the fringe, has turned into a significant portion of the farming world.

Organic food: the basics

Organic agriculture is a system of farming based on principles of human, animal and environmental health. According to the *Compendium of UK Organic Standards*, it seeks to create "optimum quantities of food of good nutritional quality by using management practices which aim to avoid the use of agrochemical inputs and which minimize damage to the environment and wildlife."

The organic concept has been around for more than half a century (the phrase was coined by Walter Northbourne in a book published in 1940), though it wasn't until the 1990s that certified organic products took off in a big way in the UK. Today, the global organic market is worth over $47 billion and supplied by more than thirty million hectares of certified agricultural land – an area roughly the size of Italy.

Since the EC passed legislation in 1993, the term "organic" has been governed by strict legal definitions and regulations across the European Union. Any product labelled as organic – wherever it was produced – must have been grown or raised according to a set of minimum EU standards and possibly extra standards imposed by the individual country and certification body.

As most people are aware, at the core of the organic movement is a policy that shuns synthetic fertilizers, pesticides, herbicides, fungicides and other man-made "inputs". Instead of these, organic farmers rely on a mixture of special growing techniques and "natural" alternatives. In place of chemical fertilizers, for example, soil fertility is maintained through the use of animal manure and crop rotation.

Crop rotation is a farming technique in which the same field is used for growing different crops in successive months or years, the idea being to ensure that the nutrients some crops take from the earth (most notably nitrogen) are put back by others. It can also reduce the need for pesticides and herbicides, though other techniques are usually also necessary. These range from manual weeding to the introduction of predatory insects to a crop (to eat the pests) or the use of "natural" pesticides such as Bt, derived from a bacterium.

Organic rules also ban additives such as artificial sweeteners, colourings, preservatives and flavour enhancers as well as hydrogenated fats and genetically modified ingredients. But this doesn't tell us the whole story. As the box on the opposite page shows, the organic "idea" is an entire agricultural philosophy, taking in animal welfare and social justice as well as environmental and health concerns.

In the case of prepared foods, to bear an organic label they must contain at least 95% organic ingredients; the rest, where applicable, may be non-organic, but only from a list of approved ingredients.

Organic enforcement

Any producer, importer, packager or processor of organic food has to register with an official certification body, which will inspect them at least once a year. In the UK, there are currently ten certification bodies, all of which impose minimum standards defined by the UK government,

The organic charter

Though the precise details of organic rules vary around the world, a definitive set of principles was agreed by the International Federation of Organic Agricultural Movements (www.ifoam.org). Here are the four principles, followed by a bit of further explanation about each, drawing on the official IFOAM commentaries.

▶ The principle of health To sustain and enhance the health of soil, plant, animal, human and planet as one and indivisible

▶ The principle of ecology To be based on living ecological systems and cycles, work with them, emulate them and help sustain them

▶ The principle of fairness To build on relationships that ensure fairness with regard to the common environment and life opportunities

▶ The principle of care To be managed in a precautionary and responsible manner to protect the health and well-being of current and future generations and the environment

The first principle "points out that the health of individuals and communities cannot be separated from the health of ecosystems" and states that "the role of organic agriculture, whether in farming, processing, distribution or consumption, is to sustain and enhance the health of ecosystems and organisms from the smallest in the soil to human beings".

The second principle "states that production is to be based on ecological processes and recycling ... and should fit the cycles and ecological balances in nature". It also reflects the organic movement's desire to reduce inputs and energy consumption "in order to maintain and improve environmental quality and conserve resources" and "to protect and benefit the common environment including landscapes, climate, habitats, biodiversity, air and water".

The third principle deals with the lesser-known aspects of organic farming: social justice ("those involved in organic agriculture should conduct human relationships in a manner that ensures fairness at all levels and to all parties") and the humane treatment of animals (which "should be provided with the conditions and opportunities of life that accord with their physiology, natural behaviour and well-being"). The idea of fairness also applies to preserving resources for future generations.

The final principle reflects the ideas that organic agriculture should reduce risk by adopting "appropriate technologies and rejecting unpredictable ones, such as genetic engineering", and that increased efficiency and productivity should never come at the cost of risks to health and well-being. It also gives the official organic line on science, which is viewed as "necessary to ensure that organic agriculture is healthy, safe and ecologically sound. However, scientific knowledge alone is not sufficient. Practical experience, accumulated wisdom and traditional and indigenous knowledge offer valid solutions, tested by time."

and some of which add extra rules of their own. The best known of the ten, the Soil Association, is often described as having the most comprehensive rules – not just among UK certifiers but globally, too.

Every organic product sold specifies the certifying body. Even if you don't see the name and logo, you'll see a code written in the form "Organic Certification UK3". For a list of the bodies and their numbers, see:

DEFRA www.defra.gov.uk/farm/organic

Generally there is a high level of trust that the organic certification system delivers on its promises. Investigative journalists and writers have occasionally found cases of rules being broken, but this isn't a major concern.

Is organic food better for the environment?

There's no single, definitive answer to this question, because environmental impact can be measured in so many different ways, and because the relevant research is a long way from comprehensive. However, according to most nonpartisan summaries of the scientific evidence – including, for example, one by staff writers at *Nature* and another by the UK government – organic farms are certainly more environmentally friendly in many ways. Compared with conventional farms, they tend to encourage greater biodiversity, such as insects, birds and other wildlife. They also tend to create less global-warming CO_2 per kilo of food (though the differences aren't huge and, as we'll see in the following chapter, there are exceptions.) Organic farms tend to generate less waste, too. In some areas, such as phosphorous run-off into streams and the all-important question of soil health, a lack of long-term comparative research makes the benefits difficult to prove beyond doubt, but, according to *Nature*, "many studies" suggest that organic production lives up to its promises.

The United Nations Food and Agriculture Organization (FAO) seems even more convinced, claiming in 2003 that: "If organic agriculture is given the consideration it merits, it has the potential to transform agriculture as the main tool for nature conservation. Reconciling biodiversity conservation and food production depends upon a societal commitment to supporting organic agriculture." The FAO also noted that organic farming "encourages both spatial and temporal biodiversity … conserves soil and water resources and builds soil organic matter and biological processes".

That's not to say that organic is necessarily the most eco-friendly farming system in the world – for now or for the future. Some scientists advocate a middle ground that builds on organic concepts but doesn't rule out all synthetic inputs or GM processes – some of which, they claim, are less harmful (both to the environment and to health) than the "natural" alternatives. Still, this middle ground isn't something that's being widely adopted and it's not something that we're offered in the shops. For now, the choice is basically between the produce of "conventional" or organic farms, and the latter are indeed better environmentally.

But even if organic *farms* are a good thing for the environment, that doesn't necessarily mean the same can be said of the organic food we buy. At least, not according to two arguments often made by critics of the organic movement. The first is that a huge amount of organic produce is flown or shipped into the UK from the other side of the world. This contributes unnecessarily to climate change, via the CO_2 emissions of the planes and boats that transport it. It's a fair point – Europe and North America account for practically all organic food sales, but only a third of total organic production, the rest being imported from Asia, Australia and Latin America. But it's also perhaps an irrelevant point: a criticism of food

Is organic food worth the money?

For most foods, buying organic means spending much more money. And, largely because of the extra labour required, it looks likely that this will never change. That said, some organic advocates argue that the organic price premium would be much smaller were it not for the fact that so much taxpayer money is spent oiling the wheels of industrial agriculture. Writing about seemingly "cheap" food, Jules Pretty, a professor at the University of Essex, has commented "we actually pay three times for our food – once at the till in the shop, a second time through taxes [for subsidies] ... and a third time to clean up the environmental and health side effects". According to one of Pretty's calculations, just the measurable extra costs – such as dealing with the damage done to the environment and to human health by certain farming practices – amount to around £2 billion a year, and that's without the massive costs of crises like BSE and foot and mouth.

These wider issues aside, it's up to each of us to decide whether organic food is worth the extra money. Looked at purely in terms of climate change, the CO_2 savings of each pound spent on favouring organic food are likely to be much smaller than those of each pound spent on, say, adding extra insulation to your home. But organic agriculture isn't primarily about reducing emissions; it also brings other benefits such as better animal welfare and farmland biodiversity.

imports as a whole, not organic farming. If we grew more organic food at home, there'd be less need to import it.

The second argument is that organic farms tend to produce less food per hectare than their more industrial counterparts. This is a pivotal issue, because it suggests that if everyone went organic we'd either face global food shortages or be required to clear vast areas of forest to make room for extra farmland – which of course would be an environmental disaster. So could organic farming produce enough for everyone without extra land? We'll discuss that question below, but first it's worth touching on one final environmental conundrum about organic agriculture. If organic farms produce less from each field, then they require extra land which could otherwise be used to grow renewable energy crops such as wood for heating and electricity generation and switchgrass for cellulosic biofuel. Arguably, then, it might reduce emissions to employ high-productivity (but environmentally sensitive) farming to produce a mixture of food and energy crops, thereby removing the need for some fossil fuels.

Could organic farming feed the world?

No one knows exactly how much food an exclusively organic world could grow, but by most estimates it would be substantially less than the amount currently produced. There are some parts of the world where small farmers switching to organic farming might actually *increase* their yields. As Nadia Scialabba wrote in a report for the FAO, "In developing countries ... properly managed organic agriculture systems can increase agricultural productivity and restore the natural resource base." Most experts agree, however, that a global switch to organic agriculture would lead to a significant fall in production.

That said, it doesn't make sense to ask whether a farming system could "feed the world" without also asking "feed with what?" After all, much of the food the world currently produces ends up squandered in the rearing of meat. Indeed, as discussed in the following chapter, the world's rainforests are *already* being chopped down to make way for more farmland. But this isn't to grow organic carrots; the crops of choice include intensively farmed soya (the majority of which is used to feed up cattle) and palm oil (used in everything from soap to biodiesel).

On these grounds, some in the organic movement claim that if we gave up our desire for cheap, intensively farmed meat (which we would have to, since huge monocrop soya farms aren't suited to old-fashioned crop rotation), an exclusively organic world could easily feed itself. Organic advocates also point out that solving world hunger is as much to do with increasing political will and changing wasteful behaviour as it is with producing more. Even according to the cautious figures of the US Department of Agriculture, one fifth of US food ends up in the bin – enough to feed the people who starve each year around the world twice over.

Still, even if a completely organic world could feed itself now, it may struggle in the future. After all, demographers expect the global population to rise to around ten billion by 2050.

The health value of organic food

Organic foods are widely marketed as being safer and healthier, but is this true? The claims fall into two separate categories, the first of which relates to the potentially harmful effects of pesticide residues in our food. Unsurprisingly, most studies concur that organic foods carry far lower levels of these residues than conventionally grown foods. The importance of this benefit is debatable – many toxicologists believe that the health risks posed by pesticide residues are extremely low (see p.240) – but most experts agree that it is a benefit nonetheless.

Regardless of whether pesticide residues are dangerous to consumers, there's little doubt about the serious health risks associated with applying certain pesticides. Figures from the World Health Organization and the World Resources Institute suggest that there are between 3.5 and 5 million acute pesticide poisonings annually, tens of thousands of which result in death. And the impact is particularly bad in the developing world, from where a significant proportion of our produce now comes. According to one study published in the *World Health Statistics Quarterly*, 99% of pesticide fatalities occur in poor countries, despite the fact that they only account for a minority of the world's pesticide use. Workers and farmers producing organic food are unlikely to be exposed to these health risks.

The second issue is whether organic foods are more nutritious. Again, there isn't universal agreement on this, but there have been many studies showing organic food to contain higher levels of vitamin C, essential minerals, cancer-preventing phytonutrients and other beneficial things. Most notably, a major study conducted by scientists at the University of Newcastle found that certain organic food products contained up to 60% more antioxidants than equivalent non-organic items grown on adjacent farms. In short, though the evidence is still a little patchy it seems reasonable to conclude that organic food is generally somewhat healthier.

Not quite organic

Set up in 1991, LEAF (Linking Environment and Farming) is a UK farm certification scheme that aims to encourage an integrated and environmentally responsible approach to agriculture. The rules are less strict than those of organics, but this, in theory at least, means that the scheme can appeal to the majority of growers who, for financial or other reasons, haven't decided to go properly organic. So pesticides are allowed, for example, but must be used responsibly. Though it's been around for more than a decade, the leaf "marque", as they call it, is one label that you're still unlikely to come across in most food outlets. Waitrose is currently the only major retailer involved.

The Wholesome Food Association (WFA), meanwhile, has similar rules to organic certifications – including no synthetic inputs – but the system is based on trust rather than inspections and paperwork. The idea is that farmers interested in ecologically sound production and local sales can participate without the cost of official organic certification.

LEAF www.leafmarque.com
WFA www.wholesomefood.org

Food transport

Meals and miles

Europe has been importing food and drink from far and wide for millennia – tea from China, spices from India, coffee from Ethiopia. But today's globalized food markets are on a scale without historical precedent. Today, we fly in fruit from the Southern Hemisphere when it's out of season in the north, and we ship in goods that we could grow in the UK, but which can be sourced more cheaply from elsewhere.

Though only a small proportion of the total food produced in the world is traded internationally – probably around 90% is consumed in the country where it is grown – the figure is much higher in most developed countries. In the UK, the distance travelled by the food we eat is thought to have roughly doubled in the last two decades. Today, the contents of an average shopping basket of goods – including organic foods – can be the result of tens of thousands of miles' journeying, by road, sea and increasingly air.

For food grown within the UK, too, the distance from "farm to fork" is bigger than ever, not least because supermarket systems rely on everything being delivered to the shop via massive distribution centres. Since these are few and far between, long truck journeys are inevitable. One much-cited study traced vegetables on sale in a supermarket in Evesham: they were grown just up the road, but had arrived via a huge round trip taking in Hereford, Dyfed and Manchester. Some defenders of the supermarket-style system – such as Lord Haskins, appointed as "rural tsar" by Tony Blair – claim that supermarket distribution may actually be greener than millions of half-full smaller vans making shorter journeys to a larger number of local shops. Whether or not this is correct (a lack of evidence makes it difficult to say), it certainly seems true that more energy is currently used to transport our food than is necessary.

As discussed below, the most obvious problem with extra food transport is the carbon emissions. But environmental damage isn't the only criticism that local-food advocates make of our increasingly long-distance dining. In the case of shipping live animals, longer-than-necessary distances raise the likelihood not only of animal welfare abuses but also of the spreading of diseases such as foot and mouth. These can end up costing astronomical sums of money: the total bill for the UK's foot and mouth crisis of 2001–02, including lost tourism revenue, is estimated to have been around £10 billion.

In the case of fruit and vegetables, there's also the possibility of food losing nutrients en route. Studies by the Austrian Consumers Association, for example, found that "fresh" out-of-season fruit and vegetables that have undergone long-distance transport are often significantly lower in vitamins and higher in harmful nitrates than genuinely fresh ones (or, indeed, frozen ones, which are usually put in the freezer within hours of being picked).

Food miles and climate change

A government report from 2005 estimates that the transport of food consumed by British residents accounts for around seventeen million tonnes of CO_2 each year. Almost half of this is caused by trucks and vans on UK roads. The rest breaks down roughly equally into four categories: trucks and vans on foreign roads driving food from the farms, ready for export to the UK; ships bringing large volumes of food into our ports; planes flying premium products into our airports; and cars driving home from the shops.

Seventeen million tonnes of CO_2 is a small but significant proportion of national emissions, so it's understandable that green-minded consumers have become increasingly concerned about food miles. In truth, though, the distance travelled by a food product is not a good indicator of its carbon footprint.

The main problem with food miles as a measure of environmental impact is that the *mode* of transport is usually more important than the distance travelled. As we saw in chapter twelve, planes are a very climate-unfriendly mode of transport. They're typically only used for freight when the items being imported are perishable and of high value. In the food sector, this mainly means things like out-of-season berries, mangetout and fine beans. In these cases, the transport will probably represent the vast majority of the product's emissions.

However, the overwhelming majority of imported foods – around 98% – are transported not by planes but by ships and trucks, and in these cases the carbon emissions are likely to account for only a small proportion of the product's total footprint. Ships typically generate between fifty and a hundred times less CO_2 for each kilo of food being carried than a plane would do. Even trucks can be twenty times less emitting than planes.

Another reason to be cautious of food miles as a barometer of greenness is the fact that it can sometimes be greener to import things than to grow them locally. For example, local tomatoes are extremely eco-friendly in the summer. But in the winter, tomatoes can only be grown in cool countries such as the UK with the help of artificially heated greenhouses, which consume large amounts of energy. Similarly, a much-publicized study by New Zealand academics claimed to show that lamb produced in their country and exported to the UK had a lower carbon footprint than lamb raised and consumed in Britain. One reason for this is that New Zealand generates most of its electricity from renewable sources, so the electricity use of the farming, processing and packing causes fewer emissions.

For reasons like this – and the fact that some foods, such as beef, will produce lots of greenhouse emissions even if they're produced just around the corner – you can't simply look at the country of origin of a food product and decide whether it's good or bad for the climate. Instead, the most important question to ask is whether it's air-freighted. Some supermarkets now label their air-freighted fresh produce, but in many cases you have to use intelligent guesswork: if something is expensive, perishable and relatively light then it may well have been transported by air, especially if it's out of season at home. Once you've excluded air-freighted foods, you

Long-distance dining: an example

In a study called *Eating Oil*, the campaign group Sustain (www.sustainweb.org) measured the miles travelled by our foods and the energy that the transportation consumes. One of their case studies looked at a basket of imported foods you might pick up in any UK supermarket.

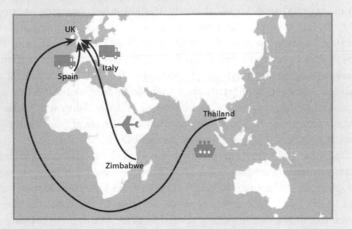

From abroad to the UK

▶ 5kg of chicken from Thailand 10,691 miles by ship

▶ 1kg of runner beans from Zambia 4912 miles by plane

▶ 2kg of carrots from Spain 1000 miles by lorry

▶ 0.5kg of mangetout from Zimbabwe 5130 miles by plane

▶ 5kg of potatoes from Italy 1521 miles by lorry

From the UK to the distribution centre

▶ 1kg of sprouts produced in Britain 125 miles by lorry

▶ All the imports to the distribution centre 625 miles by lorry

From the distribution centre to the store

▶ British sprouts, plus all the imports (totalling 13.5kg) 360 miles by lorry

Add it all up and, collectively, these goods would have travelled almost 25,000 miles, using up 52 megajoules of energy in doing so – equivalent to boiling the kettle for around 700 cups of tea. By contrast, an equivalent basket of seasonal produce from a farmers' market went only 376 miles, using up just one megajoule (around 13 cups of tea).

can further reduce the footprint of your fresh produce by favouring local foods that are in season wherever possible.

What about organic foods? These aren't likely to be any more local than non-organic foods. The UK imports the overwhelming majority of its organic produce, partly because imports are very often cheaper. Campaign group Sustain worked out that a basket of 26 imported organic products could have travelled nearly 150,000 miles, releasing as much CO_2 as "an average four bedroom household does through cooking meals over eight months". In response to these kinds of criticism, the Soil Association conducted a consultation in 2007 on whether it should cease to certify air-freighted products. In the end, the organization decided instead to limit certification to those air-freighted foods that can demonstrate fair-trade-style social benefits. This decision was taken in response to concerns about the importance of air-freight to developing world producers, an issue discussed further below.

One final point to bear in mind about food transport and climate change is that a decent slice of your food miles probably involve your own car. More than a tenth of the total emissions from food transport in, from and to the UK is accounted for by cars driving to and from the shops. If you live in the countryside and need to drive a fairly long distance to the shops, then as much as half of your food transport emissions could be the result of your driving. In cases such as these, it makes sense to try and reduce the frequency of your shopping trips (buying more each time) and to think about the efficiency of your vehicle, as discussed in chapter eleven.

What about developing-world farmers?

One obvious problem with favouring local food is the potential impact on farmers and farm workers in poor countries. Indeed, though a minority of the food and drink we consume comes from developing countries, the Fairtrade Foundation points out that "up to one and a half million livelihoods in Africa alone are estimated to be dependent upon UK consumption of agricultural and horticultural produce". If we all became strict "locavores" and shunned food from distant lands, all these jobs would be at stake.

There is a long-standing debate about the extent to which poor countries truly benefit from exporting food to the rich world. Many left-wing commentators still argue that food exports are part of a cycle that leaves poor countries overly dependent on overseas markets. Moreover, some export

National food swapping

Regardless of the debated costs and benefits of our increasingly globalized food industry, international trade in food products is unquestionably more environmentally damaging than it needs to be. Such is the weird and wonderful world of global commodity markets that countries often end up exchanging exactly the same products. Here are some examples drawn from FAOSTAT, a database maintained by the UN Food and Agricultural Organization:

▶ In 2006, the UK imported 22,000 tonnes of dried, whole milk, and exported 57,781.

▶ In the same year, the UK imported 28,508 live cows and 203,174 live pigs and exported 17,473 and 109,533 respectively.

▶ In 2005, the UK sold $665,000 of chicken to France, and imported $521,000 from the same country.

farms have been criticized over the years for their poor environment and labour standards. But leading anti-poverty groups such as Oxfam have for years argued that food exports can be pivotal to development, contributing to "rapid rural poverty reduction" in poor regions. And the jobs offered on export farms supplying Fairtrade companies or major supermarkets or food brands keen to maintain some ethical credentials are often described as the best available in those regions.

In reality, few people advocate shunning foods imported from the developing world. But many environmentalists take a strict line on air-freighted produce and – as it happens – a fairly large proportion of the food that's air-freighted into Britain comes from developing countries such as Kenya and Peru.

Oxfam points out that air-freighted foods contribute just 0.1% of UK greenhouse gas emissions while providing much-needed income in countries which have minimal responsibility for climate change but are in the front line of its impacts. On the other hand, some climate change campaigners have argued that "no emissions are sacred", and that the world must do everything in its power to reduce all avoidable emissions, whatever the short-term side effects. Some developing countries could be left almost uninhabitable by climate change, these campaigners claim, so we shouldn't kid ourselves that we're doing those countries a favour by supporting the proliferation of air-freight-dependent businesses that not only are carbon intensive but will also prove financially unsustainable as soon as tighter global controls on aviation emissions come into force.

This is one of the trickiest conundrums faced by green-minded consumers. One possible solution would be to actively favour food products from the developing world – especially those bearing a Fairtrade Mark – while at the same time avoiding the high-value perishable items that are likely to have been flown.

Whatever your view of this issue, it's important to remember that for the average consumer, transport accounts for only a small proportion of the total greenhouse impact of the food they buy – as little as a tenth, according to some figures. There are many other issues to consider, in particular the amount of meat and dairy you consume and the amount of food you waste.

GM food
Blessing or curse?

Most people in the UK had never given much thought to genetically modified foods until 1999, when scientist Árpád Pusztai caused an outcry by claiming to show that young rats fed GM potatoes were suffering ill health. Pusztai's headline-grabbing experiments are now widely accepted to have been "irrevocably flawed", as *New Scientist* put it, though some greens claim that he was the target of a shameless smear campaign and that his work still stands. Whatever the truth, the ensuing health scare about "frankenfoods" provided a springboard into the media and public consciousness for other related issues, such as GM food's ethical, environmental and economic implications.

By the time the government staged a public debate on the subject in 2003, the overwhelming majority of Britons were dead against biotechnology for food, and the big supermarkets had stopped selling most GM products. Despite this, the first UK licence for the commercial cultivation of a GM crop was given out just a few months later. In the event, the licensee – Bayer CropScience – decided to abandon the UK as a potential market, which means that no GM crops are likely to be commercially grown in Britain for some years to come. But modified ingredients and animal feed continue to be imported to, and sold in, the UK, and the arguments have broadened from questions over the safety and ethics of "tampering with nature" to claims that, by shunning GM crops, Western shoppers are inadvertently harming both the environment and people in the developing world.

The basics

Genetic modification is a kind of biotechnology in which the DNA of an organism, most commonly a crop, is altered. This can be done either by changing an existing part of the DNA or by adding a new gene from elsewhere – usually from a bacteria, a virus or another plant – allowing scientists to "cross" two organisms that couldn't combine in nature. GM has been a theoretical possibility ever since the discovery of DNA in 1953, but it was in the early 1980s that the techniques were actually developed, and in the 1990s that GM foods became a commercial reality.

Though the potential applications are very wide-ranging, there are only a handful of commercially available GM food crops at the time of writing, including soya, maize, cotton and canola. These modified plants are engineered to be either herbicide tolerant, insect resistant or both. Herbicide tolerant crops are capable of dealing with special weed killers that would kill normal crops. This, in theory at least, makes farming easier, as it enables simpler and more effective control of weeds. Insect resistant crops, by contrast, have been modified to produce a naturally occurring insecticide known as Bt toxin. This approach seeks to reduce the amount of money and time spent on buying and applying pesticides.

A public outcry has temporarily halted GM planting in Europe, but there's been no such hold-up in much of the rest of the world. At the time of writing there are around six million farmers growing GM crops on roughly 125 million hectares (approximately five times the land space of the UK). The vast majority of these are in the Americas, in particular the US, Argentina, Brazil and Canada.

Health concerns

There is nothing concrete to suggest that the current generation of GM products is more harmful or less healthy than any other industrially farmed food, and the American population, as the biotech industry is always keen to remind us, has been eating GM crops for years with no measurable side effects. However, critics such as Michael Meacher – the environment minister sacked by Tony Blair after expressing scepticism about GM – claim that this isn't enough proof of their safety, pointing out that there have been no proper "human feeding trials", in the US or elsewhere, to rigorously screen the health of GM consumers.

There have been some recent GM health-risk controversies. For example, in 2005, Australian scientists abandoned their GM pea research after the modified vegetable was found to harm the lungs of mice. There's no such evidence about GM crops on sale today, but not all scientists are convinced that the current laws, which require biotech firms to show that their crops are "substantially equivalent" to their non-GM counterparts, are enough to rule out potential increases in the levels of plant toxins or reductions in the levels of nutrients. And the tests that have been done have been largely carried out by the GM companies themselves, which renders them invalid in many eyes.

Overall, the scientific consensus, to the extent that there is one, seems to be that the risk of serious damage to human health is very low. But if a danger were found at some point down the road – by which time the GM crop in question may have widely cross-pollinated with its conventional cousins – the problem could be very difficult to undo.

Environmental impact

Environmental groups have led the campaign against GM, but biotech advocates claim that these green campaigners are shooting themselves in the foot, since GM technology could offer considerable environmental benefits. Their claims are numerous. Insect-resistant plants – with "natural" pest resistance built in – can mean less pesticide being pumped into fields, hence improving soil fertility and reducing both pollution and poisoning of farmers. Herbicide-resistant crops, meanwhile, allow for special "designer" weedkillers that are safe to use and quick to break down in the soil, and which make it easier for farmers to adopt non-tillage (plough-free) techniques, which reduce soil erosion and degradation and can help lock carbon in the soils.

GM could also make possible higher-yield crops, or ones able to grow in soil that's been left saline by irrigation (pumping water onto crops often results in salty groundwater rising up and damaging the soil). Both these technologies could help the world meet its ever-increasing demands for food without having to cut down forests to claim new farmland. And the applications aren't limited to food: biotechnology is already being used to research ecologically sound replacements for products such as bleach and formaldehyde.

Most environmentalists are unconvinced by these claims. After all, in the case of herbicide-resistant crops, there's as much evidence to suggest they end up increasing weedkiller use and harming wildlife as there is to suggest the opposite. In parts of Argentina, for example, the development of weeds resistant to the designer herbicides has led to huge amounts of other herbicides being used, harming local people's health and crops. There's also a concern that various herbicide-resistant GM crops might cross-pollinate with each other, resulting in "super-weeds" resistant to multiple herbicides. Even the British government's farm-scale evaluations of four GM varieties suggested that three could be harmful to wildlife. And that was comparing them to their equivalents from conventional intensive farms; a more meaningful study would compare GM crops to organically grown produce, environmentalists claim.

As for increasing the world's food supply, that's also a red herring, the greens argue. Most GM crops are simply supplying feed for the intensive livestock industry, the land for which is now one of the most important drivers of the disastrous clearance of Amazon rainforest. Hence we should be looking to consume less meat and redistribute more, not simply increase production.

There is one area where some GM critics concede that biotechnology may have proved environmentally beneficial. Pest-resistant crops – such as cotton plants engineered to produce Bt toxin, a pesticide based on a naturally occurring bacterium and widely used by organic farmers – seem to have succeeded in increasing yields and reducing pesticide use in China and elsewhere. But the question remains as to whether we should release any GM crops into the wild without absolute certainty that they won't cause environmental or health problems. After all, contamination (or "cross-pollination", depending on who you're talking to) is inevitable, as farmers in the US and Canada have already found. And even if this doesn't prove dangerous, it limits consumer choice and risks the livelihoods of organic farmers, since no GM-tainted crop can be sold as organic.

Some GM advocates, such as molecular biologist Conrad Lichtenstein, claim that organic farmers should simply embrace biotechnology: "GM technology … is by definition a very organic technology", he once wrote. "There is no contradiction between organic and GM." Most organic farmers and consumers beg to differ.

GM and the developing world

No part of the GM debate is as heated as the claims about the potential benefits for the world's poor. The biotech industry talks of high-yield crops able to resist harsh winters or droughts, which could reduce hunger and increase food security in regions afflicted with problematic climates and soils. And special nutritionally improved staples have already been designed to help provide much-needed vitamins and nutrients.

Anti-GM food protesters ripping up biotech crops in Banbury, Oxfordshire
Photo: Adrian Arbib

These wonder-crops have so far failed to live up to their creators' promises. So-called golden rice, for example, aimed at combating the vitamin A deficiency that blinds and kills thousands of children in poor countries every week, needs to be consumed in unrealistically vast quantities to have the desired effect. Nonetheless, these are first-generation technologies and by shunning biotechnology as consumers, we risk being the "GM Jeremiahs" – in the words of *Observer* columnist Nick Cohen – who get in the way of the development of genuinely life-saving second-generation GM crops.

This is a contentious case, made all the more so by the fact that many leaders of poor countries, as well as most global anti-poverty groups, have spoken out against GM. This resistance goes right back to 1998, when a group of African delegates to the UN issued a joint statement saying that they "strongly object that the image of the poor and hungry from our countries is being used by giant multinational corporations to push a technology that is neither safe, environmentally friendly nor economically beneficial to us". That view hasn't changed a great deal in many African countries. According to ActionAid, that's because current GM crops are "largely irrelevant for the poorest farming communities – only 1% of GM research is aimed at crops used by poor people – and they may pose a threat to their livelihoods".

Even when pro-poor GM crops do evolve, some argue they are failing to deal with the underlying problems. In *So Shall We Reap*, for example, Colin Tudge writes that vitamin A rice is the "heroic, Western high-tech solution to the disaster that Western commercial high-tech has itself created", pointing out that vitamin A deficiency doesn't exist "so long as people have horticulture", which has always been part of "traditional farming". The problem isn't low-tech rice, he claims, but a global agricultural ideology based on "obsessive monoculture" and global trade.

But the main concern of ActionAid is that "four multinationals dominate GM technology – giving them unprecedented control over their GM seeds and the chemicals that go with them". The companies in question are Monsanto, Syngenta, Bayer CropScience and Dupont. Like many other GM sceptics, ActionAid worry that these companies' virtual monopoly on seed distribution, combined with a lack of education in some countries about the legal implications of using GM seeds (for example, the requirement to buy new seeds each year), allows them to effectively force their products on farmers, who will then find themselves taking on avoidable debt.

Moreover, the big companies are registering ever more patents, giving them the sole rights to perform certain types of genetic manipulation on certain plants. So even if public-sector organizations stumble across potentially positive GM crop developments, they may find them all wrapped up in legal red tape.

Despite all this, many prominent scientists – including David King, previously the UK chief scientific advisor and an outspoken advocate of radical action on climate change – argue that Western unease with GM is holding back potentially huge benefits for the developing world. King is one of many GM advocates who have voiced this argument loudly since the food price spikes of 2008.

Whatever the truth, some farmers are clearly already benefiting from GM technologies – or thousands of them wouldn't be voluntarily choosing biotech crops in Argentina, China, India and elsewhere. And, in the longer term, there is no obvious reason why GM shouldn't be part of a solution for a better world. Certainly the UN Food and Agriculture Organization thinks so: "biotechnology is capable of benefiting small, resource-poor farmers" was one conclusion of a recent *State of Food & Agriculture* report. But whether the world *needs* GM crops, whether they're part of the solution or the problem, and whether it's the industry or consumers who are holding up any potential positive benefits, is questionable. After all, if crops truly beneficial to the world's poor are proved to be effective and safe, and offered at reasonable prices and terms, it will surely take more than a few European fears about modified cornflakes to stop them being taken up widely around the world.

GM foods: what's in the shops?

At the time of writing, no GM crops are commercially grown in the UK. And since Bayer CropScience decided to stop pursuing its goal of commercializing its Chardon LL modified maize, it looks unlikely that there will be any British GM cultivation in the immediate future except for the occasional government-approved trial (such as the one currently underway for a blight-resistant modified potato). Genetically modified crops are still imported from abroad, however, both as ingredients and as animal feed.

After a tightening of EU law in April 2004, all food derived from a GM organism must be labelled as such. This includes any product that contains individual ingredients with a GM content of 0.9% or higher. So if a biscuit was 1% soya oil, and the soya oil was 1.5% GM in origin, a label would be required. If the soya oil, for "accidental or technically unavoidable" reasons, was 0.5% GM, no label would be needed. In practice, however, this is all academic, since the major UK supermarkets, and many of the major food processors, have imposed their own stricter standards, and don't sell anything that would come close to requiring a label. The Co-op, for example, announced a number of years ago that it would stop selling any food that contained ingredients consisting of more than 0.1% GM-derived produce.

The major exception to the EU's strict GM labelling rules relates to animal products. Despite the fact that most biotech crops are grown as animal feed, there's no requirement to label meat, fish, eggs or dairy products from farm animals fed GM crops. There isn't any evidence to suggest that modified DNA can be passed from animal feed into meat or milk, but anti-GM campaigners such as Greenpeace point out that this hole in the labelling laws makes it impossible for consumers to boycott biotech foods completely. Marks & Spencer is the only major supermarket to have banned GM animal feed for its fresh meat and milk (the policy doesn't extend to frozen and pre-prepared products).

Another loophole is that labels are not required on food produced with the aid of, but not actually containing, GM-derived substances. For example, the chymosin enzyme – "rennet" – used to separate the curds and whey in cheese production is usually produced with modified yeast or bacteria. The main exception is traditional cheese made with rennet from calves' stomachs.

These loopholes aside, most people in the UK aren't currently buying or consuming GM foods in any serious quantities. If you'd rather be consuming none at all, including products from animals fed on GM crops, buy organic, since the organic rules ban all use of GM technologies. Even this may not completely remove all traces of modification from your diet, since accidental contamination of organic food by GM crops has been widely spotted. According to a study published by Mark Partridge and Denis J. Murphy in the *British Food Journal*, out of 25 "organic/health foods containing soya beans ... ten tested positively for the presence of GM material"; eight of these were labelled organic.

Fair trade
Green and beyond

The core aim of the fair trade movement is to give poor and marginalized farmers and farm labourers in the developing world the chance to improve their living standards. It isn't a "green" idea as such but, as we'll see, the fair trade system does have a surprisingly broad environmental component.

As most people understand it, the basis of fair trade is the idea that Western brands pay a minimum "fair price" for crops and labour. However, the fair trade system also aims to empower producers by – among other things – encouraging farmers to form democratically run cooperatives. In addition, the system involves making upfront payments and committing to long-term trading arrangements, to save poor producers relying on potentially crippling loans and to enable them to plan ahead.

The Fairtrade Mark

The concept of fair trade has been around much longer than the formal labelling and certification system. But today, this "official" system – which first emerged in Holland in the 1980s, in response to plummeting international coffee prices – dominates the fair-trade world. Any product bearing the label, or Mark, has been traded according to a set of internationally agreed standards and the supply chain has been audited to make sure that the rules are being stuck to. It's important to note that Fairtrade is not a brand or a company, but a certification system that farms, cooperatives, importers and brands can sign up to.

The key objectives of the Fairtrade standards are as follows:

▶ ensure a guaranteed minimum price agreed with producers

▶ provide an additional premium that can be invested in projects that enhance social, economic and environmental development

▶ enable pre-financing for producers who require it

▶ emphasize the idea of partnership between trade partners

▶ facilitate mutually beneficial long-term trading relationships

▶ set clear minimum and progressive criteria to ensure that the conditions for the production and trade of a product are socially and economically fair and environmentally responsible

These aims are achieved through two underlying sets of Fairtrade standards. One covers crops such as coffee and cocoa, which are grown by small-scale independent producers. This set of standards is primarily concerned with ensuring that farmers in "democratic and participative" cooperatives receive a decent price for their crops, rather than being left to middlemen and fluctuating global commodity markets. The other set of standards is for crops such as tea, which are produced on estates; it focuses mainly on issues such as the pay and conditions of the workers, the right to form unions, health and safety, child labour, and so on.

Besides the minimum trading requirements, Fairtrade bodies also have a set of so-called "progress requirements" – improvements in environmental sustainability, social provision and working conditions that need to be demonstrated over time. Some of these only kick in once a producer has received a certain amount of Fairtrade income.

From its niche origins, the Fairtrade Mark has become part of mainstream British culture, with more than two thirds of adults now familiar with the label. Since the first product bearing the Mark was launched in the UK – Green & Black's Maya Gold organic chocolate, in 1994 – sales of Fairtrade-labelled retail products have climbed to £712 million a year and continue to rise steeply. In the case of roast and ground coffee, they account for more than a fifth of the total market.

Fairtrade and the environment

Most people think of the Fairtrade Mark as representing social justice rather than environmental sustainability, but in fact the standards include a decent number of rules designed to ensure greener production. At a minimum, producer organizations must develop an environmental management plan, prohibit the use of certain pesticides, implement soil erosion strategies and ban all GM farming techniques. In addition, there are progress requirements for continual improvement on issues ranging from waste management and the use of fossil-fuel energy and chemical fertilizers through to the protection of virgin forest. These rules are fairly impressive in themselves, though the Fairtrade labelling organizations also "encourage small producers to work towards organic practices where socially and economically practical."

Of course, most Fairtrade goods will need to be transported a considerable distance from the producer (usually in the tropics) to the end consumer in Europe, North America or elsewhere. In the vast majority of cases – for non-perishable products such as tea, coffee, bananas and dried fruits – this transport will be provided by ships, and the associated carbon dioxide emissions will be relatively low. Air-freight may be used for some high-value Fairtrade perishables, however. In these cases, the green-minded consumer has to balance the social benefits against the high levels of carbon emissions.

Who's in charge?

After having grown independently in different countries, the Fairtrade certification system became internationally unified in the late 1990s, with the establishment of Fairtrade Labelling Organizations International (FLO). This body is now responsible for defining the standards and certifying that accredited products really are produced in accordance with them. To do this, it works "with a network of independent inspectors that regularly visit all producer organizations", and implements a trade auditing system which "checks that every Fairtrade-labelled product sold to a consumer has indeed been produced by a certified producer organization". (As with organic farming, there have been one or two cases of investigative writers uncovering shady dealings in fair-trade supply chains, but the system as a whole is widely considered to be highly robust and trustworthy.)

Based in Bonn – reflecting the fact that Germany is very much at the heart of ethical consumerism – FLO comprises a membership of 24 so-called National Initiatives that implement the system at the country level. The UK's body is the London-based Fairtrade Foundation, originally set up by CAFOD, Christian Aid, Oxfam, Traidcraft Exchange and the World Development Movement, with a little help later on from the Women's Institute. For more information on these organizations, the standards or other aspects of fair trade, see:

FLO International www.fairtrade.net
Fairtrade Foundation www.fairtrade.org.uk

Meat, dairy & eggs

Cows, pigs and poultry

Compared to air-freighted fruit and vegetables, the environmental impact of our meat and dairy consumption gets relatively little attention. That seems strange when you consider that meat and dairy products are among the most carbon-intensive foods going. According to some estimates, producing a single kilogram of beef can result in as much as 35kg of CO_2 – equivalent to driving a typical car almost two hundred miles.

So what makes some meats so environmentally problematic? There are two key issues: the amount of energy and animal feed required to make each kilo of meat, and the amount of globe-warming methane and nitrous oxide that certain animals produce, either directly or indirectly. This chapter discusses these two issues before briefly touching on each category of animal product in turn.

The background
Why eating animals is (usually) bad for the planet

The main environmental problem with animal products is that most of them are inherently inefficient to produce: it takes a relatively large amount of energy, feed and water to make a relatively small amount of meat or milk. Producing a single kilo of intensively farmed beef, for example, can take around ten kilos of feed grain. Though beef is perhaps the worst culprit, almost all meats, as well as farmed fish, dairy and eggs, raise the same basic problem. Of course, animal products have relatively high energy and protein levels, so a kilo of feed is not directly comparable to a kilo of meat. But even taking this into account, the simple fact remains: with most farm animals we get relatively few calories out compared to the food and energy we put in.

Animal products can be made far more efficiently on mixed, closed-system farms of the type often favoured by small-scale or organic growers. Chickens, for instance, can be fed partly on scraps that might otherwise go to waste. Equally, some grazing animals have a diet consisting primarily of grass grown on land that would be no good for crops. For these reasons, a global agricultural sector truly optimized for energy efficiency probably *would* produce some animal products. But in the world as it stands today, livestock farming consumes vast quantities of crops that could be fed directly to people – or indeed turned into fuels.

This issue is getting more pressing each year. Global meat consumption has rocketed by around 500% since 1950 and there is no sign of the rise slowing down. The world today contains around twice as many chickens as humans, plus 1 billion pigs, roughly 1.3 billion cows and 1.8 billion sheep and goats. According to agricultural commentator Colin Tudge, if the current growth rate of animal farming continues, by 2050, livestock will consume as much farmed feed as the entire human population did in 1970.

The rising demand for animal feed has a number of implications. First, producing it takes a great deal of fossil fuels – for making fertilizer and pesticides, powering farm vehicles and transporting the final crop to market. This fuel use causes CO_2 emissions and the application of the fertilizer causes nitrous oxide emissions, too. Secondly, and just as importantly, the demand for farmland to grow feed crops and graze cattle is already driving the destruction of rainforest in the Amazon and elsewhere.

The pressure placed on global agriculture by the demand for animal feed also has serious implications for the food security of the world's poorest people. The sharp spikes in food prices in 2008 – which led to an increase in malnutrition in many countries, and food riots in others – were caused by a number of factors, including the conversion of food crops into biofuels. But increasing global meat consumption unquestionably played an important role. As meat consumption continues to rise in the coming years and decades, the likely effect will be to push food prices upwards, making life increasingly difficult for the urban poor in the developing world.

Making meat

The exact quantity of feed required to make each kilogram of meat or fish varies according to breed and farming method. The following middle-of-the-road estimates come from the US's Council for Agricultural Science and Technology. In the UK, the figure for beef would likely be lower due to a higher proportion of free-range grazing.

Kilos of feed required to produce one kilo of food

Farmed fish	1.5–2.0	
Poultry meat	2.1–3.0	
Pork	4.0–5.5	
Beef	10	

The amount of water required to produce certain meats is also high. Of course, this is less significant in wet countries than it is in drier ones, and the figures vary enormously according to whether or not you include all the "virtual" water in the supply chain (such as the rainwater required to grow the grass or feed). According to a study by David Pimentel, published in *Bioscience*, the *total* water input for various foodstuffs grown in the US are as follows:

Litres of water required to produce one kilo of food

Potatoes	500
Wheat	900
Maize	1400
Rice	1910
Soya beans	2000
Chicken	3500
Beef	100,000

... and 180 more

In terms of climate change, the key metrics are the amount of fossil-fuel energy required to make each kilogram of food and the methane and nitrous oxide emissions of the animal and its manure. One study by Adrian Williams of Cranfield University concluded that the typical carbon footprint of British-reared meats are as follows: beef, 16kg per kilo; lamb, 17kg per kilo; pork, 6kg per kilo; chicken, 5kg per kilo; eggs, 6kg per kilo; milk, 1kg per litre. Figures for other countries or farming methods change the picture, with beef often causing as much as 35kg per kilo if farmed intensively.

Methane and nitrous oxide

Cows, sheep and goats are ruminant animals. This means their digestive system relies on a process known as enteric fermentation. Bacteria inside the animal break down hard-to-digest foods – such as grass – enabling the consumption of foods that other animals can't usefully consume.

Unfortunately, one of the byproducts of enteric fermentation is the powerful greenhouse gas methane. A typical cow belches out more than 100kg of methane gas each year. That's equivalent to more than two tonnes of carbon dioxide – or two return flights from London to New York. Sheep and goats aren't quite so prolific, but then they also produce less meat or milk per animal. Estimates vary widely, but roughly speaking

Should we all be vegetarians?

The question of whether it's justifiable to kill an animal simply so we can have the pleasure of eating it must be one of the longest-running of all ethical con-sumerism debates. Exclusively non-animal diets have been relatively common in Asia since the development of ancient religions such as Jainism, and Europe first grappled with the concept just as long ago. Pythagoras, Plato, Socrates and Ovid all expressed doubts about the ethics of meat. Western vegetarian-ism as we know it today, however, didn't really take off until the beginning of the nineteenth century. The tireless campaigning of the appropriately named William Cowherd, a reverend from Salford, started the ball rolling and the term itself was coined in 1847 by the newly formed Vegetarian Society. (The society made a point of deriving the phrase from *vegetus*, Latin for lively, rather than vegetable, though this seems somewhat academic considering that both words come from the same Latin root.) Numbers have grown ever since, and today 5% of the adult population in the UK describe themselves as vegetarian, with many more consciously limiting their meat intake.

Some people choose vegetarianism on the simple ethical grounds that they don't want to commission the deaths of animals. Others, however, choose a meat free diet in order to try and be healthier or more environmentally friendly. So are these reasons valid? Overall, yes, vegetarian diets tend to be somewhat healthier and greener, though the extent to which this is true depends on the details of the diets being compared.

From a health perspective, the scientific literature shows fairly clearly that veg-etarians have a significantly reduced chance of ischemic heart disease, certain cancers and some other health problems. But a vegetarian diet consisting pri-marily of pizza, chips and readymeals may be thoroughly unhealthy – and cer-tainly less healthy than a well balanced diet containing some meat. Similarly, a vegetarian diet will tend to be greener and more animal-friendly than a

the methane emissions of a cow or sheep roughly double the carbon footprint of the resulting meat.

Scientists have developed pills to reduce the flatulence of cattle, but these drugs haven't yet been widely rolled out and they only reduce rather than stop the methane emissions. Another approach that's been mooted is keeping cows in sealed sheds and extracting the methane from the air. But even if this works, it raises obvious animal-welfare concerns and forces the cattle to eat farmed feed rather than grazing on grass.

Even non-ruminant animals such as pigs produce methane via their manure. While manure adds to soil fertility in mixed, closed-system farms, it can be a major polluter in intensive farms – not just a source of

non-vegetarian diet but the rule won't always be true. Milk and cheese have a very high carbon footprint and, as we'll see, egg-laying chickens and dairy cows often face animal-welfare problems at least as serious as those faced by animals reared for meat. In addition, male chicks from egg-laying stock are killed at birth, while dairy products rely on a constant stream of calves being born, most of which are slaughtered for meat to avoid an ever-growing bovine population. So vegetarians still contribute to the deaths of animals, you might argue, and intensively farmed milk and eggs are not environmentally or morally superior to, for example, organic chicken. Indeed, according to one American study coauthored by Gidon Eshel at Bard College, a diet with protein from chicken has a lower carbon footprint than one featuring cheese and eggs.

These contradictions could lead you to decide not to be a vegetarian as such, but to simply limit your intake of animal products. Another response would be to go vegan, and abstain from dairy and eggs as well as meat. Veganism grew out of the vegetarian movement in the first half of the twentieth century, and it generally sees itself as the logical conclusion of that movement (as is implied by the term "vegan": the beginning and the end of the word vegetarian). It is a much bigger commitment than vegetarianism, ruling out many foods and guaranteeing a considerable degree of inconvenience when eating out or travelling. But it's ultimately a more coherent ideology, at least from the ecological and animal-welfare perspectives.

Even veganism isn't enough for some, who ask why we should draw the line at plants. Fruitarians only eat fruits, berries, olives, tomatoes, nuts and other foods that can be eaten without deliberately harming any organism, and which save energy by eliminating the energy needed for cooking, refrigeration or even washing up. This is one branch of green consumerism that it's hard to see ever really taking off.

methane but also a threat to rivers, where manure can boost nitrogen and phosphorous levels and kill various aquatic species. The ammonia content of manure can also contribute to acid rain.

The main problem with manure, however, is that it gives off nitrous oxide, a greenhouse gas around three hundred times more powerful than carbon dioxide. According to the UN Food and Agriculture Organization, livestock farming accounts for almost two thirds of man-made nitrous oxide emissions, with manure easily the biggest source. All livestock manure gives off a certain amount of nitrous oxide, but the amount is likely to be far greater if the animals have been fed on crops grown with synthetic nitrogen fertilizer. This is another reason why intensively reared beef can be so hugely polluting.

New Zealand gives an impression of how significant methane and nitrous oxide emissions from livestock farming can be. Roughly half of the country's total carbon footprint is down to these two gases being emitted from New Zealand's (mainly pastoral) agricultural sector.

Total impact and solutions

In 2006, the UN's Food and Agriculture Organization published a study examining the total contribution of livestock farming to environmental problems such as climate change. The report concluded that if you factor in deforestation for grazing and feed-growing land, plus the fossil fuels used on the farms and the methane and nitrous oxide emissions from the animals and their manure, then red meat (and leather) production accounts for a staggering 18% of the entire human impact on the climate each year – substantially more than the global transport sector.

Various groups claim to have solutions to reduce the environmental burden of our animal-derived foods. Carbon footprint experts encourage us to switch from beef and lamb to pork, chicken and eggs; vegans advocate simply cutting out all animal products from our diets; GM exponents look to biotechnology as a means of increasing feed production and decreasing methane emissions; and some food economists have even suggested the implementation of a global "food-chain tax", to add to the cost of meat and disincentivize its consumption.

Perhaps the most intriguing proposal for greener meat is to use special cattle farming techniques to help lock up extra carbon in damaged soils. This approach, discussed in Chris Goodall's book *Ten Technologies to Save the Planet*, involves grazing practices that mimic the movement of wild herd animals such as buffalo. The system's proponents claim that this kind of farming can reverse desertification, encourage healthy plant growth and in doing so remove large volumes of CO_2 from the air. It remains to be seen whether this kind of farming will be recognized as a carbon sequestration technique in the next round of global climate talks.

In the meantime, individuals concerned about climate change, deforestation and other environmental issues would do well to reduce their intake of animal products, especially beef and lamb.

Animal welfare labels
What do they mean?

Even if meat production wasn't environmentally problematic, there are serious problems with the conditions that many animals are subjected to. Specific welfare concerns are covered in the following sections on meat, poultry, eggs, dairy and fish. But first here's a quick look at two schemes that claim to guarantee better levels of welfare. Interestingly, neither is as strict as the Soil Association's organic standard, which remains the best guarantee of good treatment of farm animals.

British Farm Standard

Launched just after the turn of the new millennium in response to growing public distrust of the agricultural sector, the NFU's Little Red Tractor scheme labels meat, dairy, veg and cereals that come from farms conforming to British Farm Standards – a set of "exacting" standards relating to the environment, food safety and animal welfare. The scheme

BRITISH FARM STANDARD

sets out some very basic minimum animal welfare standards – such as outdoor animals having access to shelter – but, according to campaign groups such as Compassion in World Farming, the rules still allow such animal-unfriendly practices as narrow farrowing pig crates, battery cages for laying hens, restrictive feeding practices, combinations of breeds and diets that can lead to animal health problems, debeaking of chickens, and tail docking and teeth clipping of pigs. Friends of the Earth have also complained that it doesn't rule out GM. For more information, see:

Red Tractor www.redtractor.org.uk

RSPCA Freedom Food

Established in 1994, the Freedom Food scheme certifies eggs, dairy, chicken, pork and some prepared foods. The idea is to label products from animals that – from farm through to abattoir – haven't been denied any of the basic "five freedoms": freedom from fear and distress, freedom from hunger and thirst, freedom from discomfort, freedom from pain, injury and disease, and freedom to express normal behaviour.

As interpreted by the scheme, these freedoms don't preclude some unpleasant practices – hens might still be debeaked and kept in high-density sheds, for example – and this, combined with the fact that farmers pay to join, has led to criticism from some quarters. While Animal Liberation Australia's claims that the RSPCA is "in bed with the egg industry … in the business of harming animals" are over the top, it's true that the scheme is less strict than you might imagine, considering who runs it. But the rules are certainly far better than nothing, and were they any tighter it might effectively limit the scheme to the relatively small premium-product marketplace rather than making improvements across mainstream farming.

For more information, or a full list of products available at each supermarket, visit the RSPCA's website. Or for stricter animal welfare standards, go organic.

RSPCA www.rspca.org.uk

Red meat
Beef, lamb & pork

Beef and veal

As we've seen, there's a strong case for avoiding beef products altogether, due to their high carbon footprints: 16–35kg per kilo of meat. If you do buy beef, however, try at least to favour meat from free-range suckler herds, which spend most of their lives moseying around in fields. This approach – which is very common in the UK, unlike much of the US – minimizes energy and feed requirements and is much more animal-friendly. The cows get to exercise properly and the calves stay with their

The antibiotics issue

Along with their diet, many animals in intensive farms – especially pigs and poultry, but also cattle and fish – routinely receive drugs. These are mostly antibiotics, which can be used both to stimulate growth and to prevent the diseases that thrive in overcrowded conditions. Animal welfare campaigners object to antibiotic growth promoters because they are thought to cause health problems as well as enabling farmers to push animals beyond their natural limits. But equally troubling is the potential threat to human health via the growth of antibiotic resistance. As "superbugs" in hospitals have shown, resistance to our most useful drugs is on the up. For instance, a significant proportion of cases of campylobacter, the most common cause of food poisoning in the UK, are now untreatable by antibiotics. Though not all scientists agree, some microbiologists think we're sitting on a time bomb.

Over-prescription in humans is certainly part of the problem – perhaps the biggest part – but intensive farming is now widely accepted as a potential breeding ground for resistant bugs, which, once developed, may be passed on to humans via meat, milk or even possibly crops grown from animal fertilizer. The chicken industry quietly went back on its much-publicized phasing out of antibiotic growth promoters, but these additives were finally made illegal across the EU at the start of 2006. Still, growth promoters only account for a small proportion of the hundreds of tonnes of veterinary antibiotics used each year, and there are no plans to phase out the rest, which are routinely administered on a prevention-is-easier-than-cure basis.

Drugging animals also raises the question of carcinogenic or otherwise harmful drug residues in our meat or milk. Strict regulations setting out periods of "withdrawal" for animals given drugs theoretically avoid this, but various studies have shown residues in excess of the legal maximums, suggesting farmers don't always observe the rules.

If you want to avoid routinely administered antibiotics, buy organic or shop at Marks & Spencer, which claims to enforce a strict policy governing the use of veterinary medicines.

mothers for up to a year. As a bonus, free-range beef is usually better quality, too.

By contrast, some cheaper beef comes from intensive farms, usually stocked with excess calves from the dairy industry. (Old worn-out milkers were also once used for meat but today, post-BSE, their carcasses are burned instead.) In these farms, the cows are kept partly or entirely indoors, often in over-crowded concrete sheds, and have little chance to exercise or graze. As with most intensive farming, this increases the risk of many illnesses, which in turn heightens the need for antibiotics and other drugs. The only benefit of intensive beef is that cows raised this way tend to emit slightly less methane, though this benefit is not sufficient to offset the carbon footprint of producing the extra feed they require.

Veal is almost always intensively farmed and often with minimal concern for animal welfare. Things aren't as bad as they used to be: battery-style veal farming, in which calves are reared in crates so small that the animals can't even turn around, have finally been phased out across the EU. Nonetheless, veal is still often produced in all-indoor farms, so it continues to pose welfare problems as well as being a high-carbon food.

Lamb & mutton

The greenhouse gas emissions involved in rearing sheep are surprisingly high, thanks to the methane produced by the animals and the nitrous oxide emissions of their manure. As mentioned earlier in this chapter, one study concluded that, on average, British lamb is even more carbon intensive than British beef, causing 17kg of CO_2 equivalent per kilo of meat. The same study found that organic lamb offered a significant improvement – 42% less greenhouse gas – which puts it somewhere between pork and beef in the carbon footprint stakes. According to another study, lamb from New Zealand has a lower carbon footprint than lamb from the UK even when the shipping emissions are taken into account (see p.194). It seems fairly likely, therefore, that organic New Zealand lamb will be the greenest version available.

In terms of animal welfare, sheep get a pretty good deal compared to most farm animals. The overwhelming majority spend their lives freely grazing over extensive pastures, being brought inside only when the weather demands it. This is largely because, as yet, no one has worked out a way to make intensive sheep farming viable. That's not to say the situation is perfect. Animal welfare campaigners claim that there are some serious problems with the way in which sheep and lambs are transported and slaughtered, and neglect by some farmers means that around one in ten lambs dies of cold or hunger. Many are also castrated without anaesthetic. To ensure the highest standards of animal welfare, opt for organically certified lamb and mutton.

Pork products

Unlike cows and sheep, pigs don't emit methane, and they're better than cattle at converting their feed into meat. Pork is still a carbon-intensive

Two extremes of pig farming: the worst of intensive and the best of organic
Photos: Compassion in World Farming

food, however, producing around eight times its own weight in CO_2. Moreover, of all the widely available meats, this one raises the most pressing animal welfare questions. Pigs are highly intelligent social animals, comparable to dogs in the IQ stakes, and yet they're often reared with little respect for their wellbeing. Pig farming techniques have become increasingly industrial over the past few decades. Farms are growing to epic sizes (in the US, there are now individual farms with more than a million swine) and the sow has been transformed into a meat-making machine. Relatively recently, female pigs would give birth to around five piglets a year, but twenty-five is not uncommon today. This is thanks to a mixture of intensive breeding practices and the technique of removing piglets from their mothers very early, to maximize the number of possible pregnancies per year.

The UK is better in terms of pig welfare than most of Europe, where many pigs still spend their whole lives in sow stalls – barren cages, often so small that the pig can't turn around for an entire pregnancy – or chained to the floor with a tether. Routine "docking" of tails and clipping of teeth is also common in much of Europe, to avoid the pigs biting each other – something they generally do only in cramped conditions with nothing to keep them entertained. Most of these techniques are due to be made illegal across the EU, but not until 2013.

In the UK, things have been improved by various bits of legislation, but the situation is far from rosy. According to Compassion in World Farming fewer than half the pigs in the UK are even provided with straw. This leaves them on bare floors, which can cause health problems as well as preventing normal behaviour. Moreover, farrowing crates – small stalls into which sows are put for a week or two either side of giving birth – are still the norm. The industry claims these are needed to protect piglets from being crushed by their mother, but organic farmers have shown that with good husbandry farrowing crates aren't necessary.

When you do buy pork products, favouring British – and preferably local – producers will help reduce the meat's transport emissions as well as ensuring somewhat better animal-welfare standards. Better still, look out for free-range pork. At minimum, you'll know the pigs aren't kept permanently indoors. Note that outdoor-reared is better than outdoor-bred – the latter ensures that the pig was *born* in free-range conditions (so no farrowing crates), but the former also implies that it was *raised* outdoors.

Organic pig meat – which is free-range but allows pigs to be kept indoors for a proportion of the time – offers further improvements. Farrowing crates are out; rooting and exercise areas are in; piglets are not removed from their mother so early; the diet is better; and medicines used must be approved by a certifying body.

Poultry & eggs

Factory fowl

Poultry meat and eggs usually have smaller carbon footprints than foods from cows and sheep. But, as with pork, poultry farming is problematic in other ways. This is a whole sector of high-tech agriculture carried out almost entirely behind closed doors, and with little respect for animal welfare. Just a few companies provide the majority of the world's chicks, and they're proud of their scientific approach. Visit the website of one of their brands – such as www.aviagen.com – and you'll be invited to check out the "features and benefits" or a "technical data sheet" for each of their "products". If this sounds more like marketing for cars than hens, the names of the breeds – such as the Hybro PG+ and Cobb 500 – won't convince you otherwise.

Some free-range but non-organic chicken and egg farms, such as this one in Rotterdam, are more densely stocked than many consumers imagine
Photo: Corbis

Chicken

With the possible exception of wild-caught game, chicken is probably the most climate-friendly meat going. That's because hens are relatively good at converting their feed into flesh. For typical UK production, chicken meat causes around 5kg of CO_2 per kilo of meat, according to one academic study. That's still far more than many pulses and vegetables but around two thirds less than beef or lamb, and slightly less than pork.

When it comes to animal welfare, however, industrially produced chicken leaves much to be desired. To keep prices low and production high (British people consume around 850 million chickens per year) farms use chicks that have been specially bred to grow in the shortest possible time. A modern broiler (meat chicken) can get to slaughter weight in just over forty days, twice as quickly as a few decades ago. Some industry insiders expect thirty days to be realistic in the near future.

Regrettably for the birds, while their breasts grow unnaturally quickly, the rest of the body cannot always keep up and, according to groups such as Compassion in World Farming and the RSPCA, millions of broilers each year develop leg disorders or die of heart failure. (The genetic selection for fast breast growth is so rigorous that chickens kept for breeding broilers have to be severely underfed; if they weren't, their super-efficient ability to put on weight would kill them before they reached egg-laying age.)

Most broilers live in flocks of thousands in giant, windowless sheds, each bird getting about as much space as a piece of A4 paper. With such dense stocking, clearing out the litter floor is usually only possible once the birds are taken to slaughter, so they sometimes have to sit in their own filth, which can lead to ulcerous feet and skin disease. Evidence of the latter was found in 2% of the chicken meat examined in a sample by *Which?* magazine.

But chicken poses an interesting dilemma for environmentally minded consumers. Industrial chicken farms – for all their obvious problems – are at least run with maximum feed-efficiency in mind. This reduces the carbon footprint of their products. According to the so-called "Shopping

How free is free-range?

Various categories of chicken and poultry with higher welfare levels are available in the UK, but the labelling isn't as simple as you might expect. Here's a rundown of the various categories of poultry meat you'll come across:

▶ Extensive indoor or barn-reared The same as standard intensive systems but with lower stocking densities and a slaughter age of at least 56 days.

▶ Free-range Basically the same as above but the birds have continuous daytime access to open-air runs for at least half their lifetime.

▶ Traditional free-range As above but with smaller flock sizes, a minimum slaughter age of 81 days, and access to open-air runs from six weeks of age at the latest.

▶ Free-range total freedom As above but with "open-air runs of unlimited area".

▶ Organic Organic standards require welfare at least as good as traditional free-range, but demand far smaller flock sizes, a better diet (mainly organic and free from animal protein), a slightly older minimum slaughter age and tighter pollution restrictions.

If neither organic nor free-range chicken is available, at least look out for the RSPCA Freedom Food label, and avoid fresh chickens with brown "hock burn" marks near the knee joints – these often imply that welfare standards have been low.

Trolley Report", commissioned by the UK government, the total carbon emissions of industrially farmed chicken (4.6kg of CO_2 per kilo of meat) are lower than those of free-range (5.5kg) and lower still when compared to organic (6.7kg). So is it better to buy cruel, low-carbon chicken or humanely produced chicken that has a higher carbon footprint? Perhaps the best answer is to buy organic but also to buy less.

Other poultry meat

Most poultry – including turkey, duck, quail and guinea fowl – are intensively farmed in a similar way to chicken. Good data isn't available on their various carbon footprints, but the animal-welfare issues are similar to those associated with chicken and the same multi-level free-range labelling systems apply. One exception is geese, which (so far at least) have not been able to survive the rigours of industrial farming, so are necessarily free-range.

Eggs

Perhaps surprisingly, eggs have a higher carbon footprint than chicken meat. That was the conclusion of the Cranfield University study mentioned earlier in this chapter, and other academic reports have concurred. In terms of emissions per kilo of food, eggs are equivalent to pork, causing about six times their own weight in CO_2.

Beyond climate change, egg farming raises the same issues and conundrums as broiler production. The RSPCA estimates that there are twenty million egg-laying chickens in the UK and that the majority of these are kept in cramped battery cages, stacked up high in vast buildings with little or no natural light. In most cases the hens have so little space that they cannot even turn around or stretch their wings, let alone follow their instincts to preen properly and dust-bathe. The cages lead not just to discomfort, but to foot injury and weak bones, so that by the end of their productive lives – usually about one year and a few hundred eggs – almost a third of the birds have broken bones, according to animal-welfare groups.

An egg-layer's diet contains not only grains, soya and nutrients, but often also fish products and even (in an odd twist on the chicken-and-egg question) waste poultry extracts. The hens are also given colorants, both artificial and natural, to ensure the consumer isn't faced with a perfectly harmless variety in yolk colour, and antibiotics are widely used. A study by the Soil Association showed that a large proportion of eggs contain traces of the toxic medicine lasalocid, known to be harmful to mammals but never tested on humans.

One advantage of cages is that they make it harder for crowded birds to peck each other, meaning that they don't necessarily need to be debeaked (the trimming of chicks' beaks with a hot blade). However, many battery hens are debeaked anyway, and most commentators agree that minimal debeaking is less cruel than battery conditions. Either way, the practice isn't necessary with a small flock size in better conditions, as is typical of organic egg production.

As with chicken meat, however, organic production may have a down side, too. Some studies suggest that the production of organic free-range eggs takes an extra 15% energy, with greenhouse gases up 20%. Once

How to read an egg box

▶ **Farm, Fresh, Country, etc** Terms like these mean nothing at all, with the exception of "Super Fresh", which relates simply to the delay between laying and packing.

▶ **The Lion Quality mark** appears on around 75% of UK eggs. The scheme was set up in the wake of the salmonella crisis, and it focuses on food safety rather than animal welfare or broader environmental practices.

▶ **Four-grain eggs** come from hens fed on natural (non-animal) food, but this doesn't reflect on the welfare of the birds.

▶ **Barn eggs** are produced in similar indoor conditions to battery eggs but without the cages. This allows the chickens to move around, but conditions are usually still overcrowded, leading to the birds attacking each other and even eating each others' droppings. Debeaking is necessary.

▶ **Freedom Food, RSPCA approved** requires basic welfare standards to be in place. The scheme rules out battery cages, but not crowded barns, colorants or debeaking – and it doesn't necessarily mean free-range.

▶ **Free-range eggs** are from hens that have had continuous access to outdoor runs "mainly covered with vegetation". However, the sheds are still usually overcrowded (one square foot per bird is permitted) and the colony sizes are large, resulting in unnatural behaviour from the birds (sometimes including, ironically, a fear of going outside). Debeaking is standard.

▶ **Organic eggs**, are from small flocks of birds that have free-range access to organic pasture and a diet consisting largely of organic food. Debeaking, routine antibiotics and colorants are banned.

again, then, the conscientious consumer needs to weigh up carbon footprint on the one hand with animal welfare and broader ecological impacts on the other.

Dairy products
Milk, cheese and butter

Many vegetarians shun meat primarily because of environmental and animal welfare concerns. But dairy products raise many of the same issues. As we've already seen, butter and cheese can have a higher carbon footprint per calorie than some meats do. One German study concluded that a block of hard Cheddar-style cheese caused 8.8 times its own weight in carbon dioxide. This large footprint is partly due to the energy used in the production of the milk (it takes around ten litres to make a kilo of hard cheese) and partly due to the fact that cows belch out methane and create nitrous oxide regardless of whether they're being bred for meat or dairy.

As for animal welfare, the milk trade not only props up the veal industry but is also sometimes responsible for animal suffering itself. Modern dairy cattle have been intensively bred to yield the maximum quantities of milk. While a calf would naturally suckle a few litres a day from its mother, milking machines now extract up to fifty litres. Indeed, breeding has been so intense in the last few decades that many cows have grown too big for older milking parlours. The extra productivity of each cow boosts economic efficiency and probably lowers the carbon footprint of the milk (since a cow that produces twice as much milk probably won't consume twice the amount of food and belch up twice the methane). The flip side is the health impact on the cattle. According to animal-welfare groups, today's cows often have oversized udders, which can create spinal problems, lameness and other damage, and a significant proportion develop mastitis, a painful udder infection.

To keep them milking, dairy cows are usually made to have calves each year or two, so for much of the time they're both pregnant and lactating. Newborn calves are usually taken from their mothers almost immediately – often within hours of birth – which according to animal-welfare campaigners causes great distress to both. Female calves typically go on to become milkers; males are usually taken straight to the market to be sold for veal or moved to intensive farms to be fattened up for low-quality beef.

As with beef, the environmental and animal welfare problems associated with dairy production vary widely according to the farming practices. In some intensive farms, herds are kept inside for far more of the year than necessary, often in cramped and uncomfortable conditions. On organic farms, by contrast, cows get a better diet and living conditions, spend more time outside and have a lower incidence of lameness. Preventative antibiotics are banned. Calves stay with their mothers for longer and aren't weaned off milk for at least three months, or removed from the farm before six months.

As with many animal products, however, the climate change benefits of organic dairy aren't as clear cut as the animal welfare benefits. Certainly organic dairy farms use less energy: the question is the extent to which this advantage is counteracted by other factors such as a grass diet leading to more methane emissions. The Cranfield University report cited elsewhere in this chapter estimates that British organic milk has a higher carbon footprint than British non-organic milk by around 20%, though other studies have reached very different conclusions.

Whatever the truth about organic and non-organic dairy products, the fact remains that – as with beef – these products will have a fairly high carbon footprint per calorie, no matter how they were produced. So the most meaningful step you can take from a global warming perspective is simply to try and consume less dairy, along with less meat. That might mean reducing portion sizes, or it might mean switching to vegetable-based alternatives – for example, substituting margarine for butter. This will certainly reduce your carbon footprint and it may make your diet healthier, too.

Fish &
seafood

The World Resources Institute estimates that fish consumption has risen more than fivefold in the last half-century. In some ways this is a positive thing. Fish provide a healthy and important protein source for hundreds of millions of people around the world, and – unlike livestock – they don't require vast areas of crop land for feed. Unfortunately, however, the massive global increase in fish consumption has been made possible by two ecologically problematic developments: industrial sea fishing and intensive fish farming. This chapter looks at the issues associated with sea-caught fish and farmed fish in turn before offering advice on sustainable choices. First of all, though, let's briefly touch on the climate impact of fish consumption.

The carbon footprint of fish

The greenhouse impact of wild-caught fish can be extremely low or extremely high depending on how and where the fish was caught. Hobby fishing must be one of the most climate-friendly ways of acquiring animal protein (unless you drive a long way to get to the river or sea). At the other end of the scale, a high-value sea fish or crustacean caught far out at sea by a heavy trawler boat and delivered to its final market by refrigerated plane might be more carbon intensive than almost any other food. The emissions of farmed fish can be similarly varied. According to the UN Food and Agriculture Organization, farmed carp is a relatively low-carbon animal product, whereas farmed salmon and shrimp have "very high" footprints.

At the time of writing, there's very little robust data about the carbon emissions of individual species and fisheries, but it is possible to reduce the carbon footprint of your fish consumption by following a few general rules. First, avoid over-fished species, as these will generally require fishing boats to travel further for each kilogram of fish. (You'll find guidance on which species are over-fished later in this chapter.) Second, avoid expensive, exotic fish: species that are indigenous to the country where you live are less likely to have been imported by air. Third, though it's not always easy to know the fishing techniques used, try to avoid fish and seafood captured by trawling. Figures published by Seas at Risk, the umbrella group for marine environment organizations, suggest that trawling is a very energy-intensive approach. For example, lobster trawling requires nine litres of fuel for each kilogram caught, compared to two litres when traps are used. A similar reduction can be made by catching cod with gill nets instead of trawls.

Sea-caught fish
Tuna, cod, swordfish et al

As fishing vessels have grown from small wooden boats to giant factory ships, the world's waters have been reaped so heavily that the proverb "plenty more fish in the sea" is looking distinctly tenuous. By the time industrial fishing came of age in the early 1980s, the global catch in two years was equivalent to all the fish caught in the nineteenth century. And today, according to UN figures, a quarter of the world's fisheries are either over-exploited or depleted, while another half are being fished to maximum safe levels. The best-known example of an over-fished species is cod – which in some areas has gone from being richly plentiful to basically extinct – but this certainly isn't the only species to have suffered.

People have long argued about the extent and implications of over-fishing, but a ground-breaking study published in *Nature* in 2003 suggested that the problem may be far worse than was previously expected. Having spent ten years scrutinizing early fishing records, Canadian academics reached the staggering conclusion that in the last half-century 90% of all large fish – including halibut, marlin, swordfish, sharks and tuna – have been wiped out. And those that are left are tiny compared with their relatively recent ancestors.

The implications could be far-reaching, as over-fishing not only knocks out the target fish but also causes havoc in the wider food chain. If these trends continue, the effect could be a "complete reorganization of ocean ecosystems, with unknown global consequences", in the words of one

Bluefin tuna – a valuable fish popular for steaks and sushi – being hauled onto a boat in Sicily. Bluefin has become an endangered species due to over-fishing.
Photo: Corbis

of the *Nature* report's authors. This is particularly concerning at a time when climate change is already threatening many marine ecosystems by increasing the temperature and acidity of the oceans.

Currently, many of the most over-fished areas are around Europe and North America. Europe has implemented a quota system to limit the damage and allow replenishment. But even if this is sufficient (which many experts doubt) and the rules are implemented properly (which so far they haven't always been), other areas are feeling the pressure in the meantime. The EU, for example, has already purchased fishing rights from a number of African countries. This may provide income for countries that need it, but some commentators predict that a lack of regulation will soon see stocks in these regions depleted, resulting in dwindling catches for local people who rely on fishing to survive.

Bycatch and sea-bed damage

Fishermen have always hauled in a certain amount of bycatch – non-target fish and other marine life. But as fishing has become more industrial, the problem has become ever more serious. The relationship between tuna fishing and dolphin deaths is relatively common knowledge (see box), but the problem actually is much wider, with turtles, sea birds, seals, whales and sharks routinely getting caught in nets, not to mention huge quantities of small fish, which are often just thrown back dead.

Driftnets, used to catch tuna, swordfish, sardines, herring, albacore and other species, are one major culprit. Despite recent legislation, these nets are often many miles long and have earned the nickname "walls of death". But bottom trawling, which entails dragging a fine net over the sea bed, can be even worse. This technique – widely used for prawns, scallops, plaice, clam, snapper and other species – can result in more bycatch than target and can also cause serious damage to the sea bed. In biologically diverse areas around underwater "seamounts", *New Scientist* recently reported, trawling is wiping out scores of species before they can even be identified.

Tuna and dolphins

The one example of the bycatch problem entering the public imagination was the 1980s campaign about dolphins being killed in tuna nets. This controversy relates primarily to yellowfin tuna, which live in the Eastern Pacific and swim in large schools underneath groups of dolphins. Since the middle of the twentieth century, fishermen have exploited this relationship: find a group of the easily visible dolphins, surround them with large purse-seine nets, and pull in the tuna underneath, usually along with the dolphins, which are thrown back dead. This practice is thought to have killed many millions of dolphins, seriously depleting their numbers, which have not recovered since. Driftnets and other tuna-fishing techniques have also been responsible for killing some dolphins.

The 1980s consumer boycott resulted in improved fishing techniques, tighter regulations in some countries, and the Dolphin Safe label from the Earth Island Institute. All this has made a big difference and though some dolphins continue to die in tuna nets only a tiny amount of the tuna on the European market now comes from so-called "dolphin-deadly" fisheries. The issues aren't quite cut and dry, however, because some dolphin-friendly fishing methods – such as attracting the fish with floating logs or using baited hooks hanging off mile-long floating lines – can have other types of adverse impacts. They may be better for dolphins but worse for sharks, turtles and non-target fish, and they may risk damaging tuna stocks by yielding younger fish that haven't yet reproduced.

Another concern is straightforward over-fishing of tuna stocks. The situation isn't too bad for skipjack and yellowfin, which dominate the tinned tuna market, but some other species – most notably the giant bluefin, which is mainly used in sushi – have been fished near to the point of extinction.

Farmed fish

Salmon, plaice, trout et al

With the seas suffering from depleted stocks, fish farming, also known as aquaculture, seems like an obvious solution. The proportion of our fish that has been farmed is rising quickly, making aquaculture one of the fastest-growing of all food sectors. Almost half the fish consumed by people around the world is now thought to have been reared rather than caught. For some specific fish – such as salmon – the figure is much higher. For years the lochs of Scotland and the fjords of Norway have been gradually filling up with the salmon industry's giant plastic nets. Farms now account for more than 99% of the Scottish salmon on sale. Trout is also widely farmed and, with stock numbers low and financial potential high, the farming of cod, sea bass and bream is on the up.

Farms certainly get around the problem of bycatch and satisfy consumer demand for cheap fish. Unfortunately, though, instead of alleviating the over-exploitation of the sea they very often exacerbate it. The problem is that to farm carnivorous fish like salmon, cod and haddock, you need to catch a huge quantity of smaller fish to feed them. For each kilo of salmon you buy in a shop, up to five kilos of fish will have been caught, transported and turned into fish food.

In addition to this wastefulness, environment groups claim that much aquaculture is environmentally damaging in other respects, too. One problem is that fish farms generate a lot of mineral-rich faeces. Scottish salmon, for example, excrete twice as much phosphorus as the country's human population, according to WWF. This goes straight into the surrounding waters, suffocating sea-bed life and possibly creating the toxic algal blooms that have left much of Scotland closed to shellfishing for many years. Fish farm waste also contains a cocktail of antibiotics, uneaten food and dyes (farmed salmon would be grey if dye wasn't added, since salmon only turn pink in the wild due to a diet rich in tiny shellfish). Various illegal chemicals and hormones, banned for their environmental and health impact, have also shown up in spot checks from time to time.

Another potential impact of fish farms is a reduction in wild stocks nearby. Sometimes this happens through the inevitable spreading of the diseases and lice that thrive in intensive farms, but there may also be a more sinister mechanism at play. Farmed salmon, for example, regularly escape in huge numbers (WWF estimates that 630,000 salmon escaped from Norwegian farms in 2002 alone) and if these fish, not adapted for life

The problems of prawns

Prawns fall into two categories. There are the small cold-water ones frequently used in sandwiches, and the large, tropical "tiger" or "king" prawns that are increasingly popular in restaurants and supermarkets. According to campaign groups such as Christian Aid and the Environmental Justice Foundation, tiger prawns are about as unethical a crustacean as you can consume, their cultivation involving the very worst practices of both farmed and caught seafood.

The majority of the tiger prawns that reach the UK have been farmed in Asian countries such as Bangladesh and the Philippines. Tropical prawn farming has a long history as a sustainable aquaculture, but the large-scale modern methods stand accused of being highly destructive. According to critics, the prawn farms' man-made pools drain local water sources, requiring an estimated 50,000 litres of water for each kilo of prawns produced, and can become a major source of pollution. Salt is added to the water, as well as fishmeal food, antibiotics to limit the risks of overcrowding, growth stimulants, and lime to regulate the acidity. This saline cocktail inevitably filters back into nearby agricultural land, rendering it unproductive, and pollutes local drinking water sources.

Nearby fishing areas also get hit, according to campaigners, both by direct pollution and by aggressive industrial fishing for potential prawn-feed (like salmon, prawns consume far more food than they produce). Moreover, the farms have been responsible for much of the widespread destruction of ecologically valuable mangrove swamps. A remarkable 25% of the mangrove forest lost each year is due to prawn farming, according to the Marine Stewardship Council. In some cases the ponds end up so toxic or virus-ridden that they have to be abandoned, their owners moving on to new locations and leaving local people with the environmental mess. There have even been reports of people being violently displaced from prospective pond sites, as well as murder and rape in some extreme cases.

In December 2004, the BBC broadcast a documentary focusing on the impacts of prawn farming in Honduras and concluded that the problems were being exaggerated by the likes of the Environmental Justice Foundation. The EJF, in turn, accused the BBC of poor journalism, not least because the film focused on a single farm. Whatever the truth, the UN was concerned enough about the shrimp industry by August 2006 to publish a set of International Principles for Responsible Shrimp Farming. It's hard to say whether these have yet brought about any major changes on the ground.

Not all tiger prawns are farmed: others are sea-caught. But these aren't necessarily better, since prawn trawlers are associated with problems of their own. The technique used – dragging a fine net over the sea bed – can damage the sea-bed as well as bringing in as much as twenty times more bycatch than prawns, endangering turtles and other sea life and diminishing the catch of subsistence fishermen. According to the EJF, prawn trawling accounts for a third of the world's bycatch but produces only a fiftieth of its seafood.

in the natural world, breed with wild salmon, a kind of negative evolution takes place, with wild fish becoming less and less able to cope with their natural conditions.

Finally there's the issue of animal welfare. Few people find it easy to empathize with fish, but keeping creatures that are naturally migratory and/or solitary in cramped spaces with 50,000 others certainly raises ethical questions. So do practices such as allowing fish to suffocate as a way of killing them, and stocking them in such high density that their fins are regularly damaged – two things reported to be common practice on some trout farms.

Sustainable fish
What to eat, what to avoid

The problems described above are serious, but they don't necessarily mean that we should stop eating fish and seafood. There are various measures you can take to minimize the negative impacts of what you buy. One option is to look out for fish products that have been certified as being sustainable by the Marine Stewardship Council (see box overleaf). Another approach is to be picky about which types of fish and seafood you choose. The quandaries of tropical prawns, farmed salmon and some large sea-caught fish have already been discussed, but there are many other species endangered by over-fishing or caught in destructive ways, and others still which are environmentally unproblematic. For an in-depth view, the Marine Conservation Society's consumer website – www.fishonline.org – provides details on the issues associated

Marine Stewardship Council

MSC-certified New Zealand hoki

The Marine Stewardship Council

Any seafood bearing the blue tick/fish label shown to the right has been caught according to sustainable criteria set out by the Marine Stewardship Council (MSC). Initially set up by Unilever and the WWF, but now an independent organization, the MSC both promotes a responsible approach to fishing and monitors practices. To be awarded the mark, fisheries need to be able to demonstrate their commitment to:

▶ The maintenance and re-establishment of healthy populations of targeted species, and the integrity of ecosystems

▶ Effective management systems, considering biological, technological, economic, social, environmental and commercial aspects

▶ Following relevant local, national and international laws, standards, understandings and agreements

After a fairly slow start, the MSC has picked up steam recently, with 51 fisheries now certified. These include pollock and salmon from Alaska; hoki from New Zealand; hake from South Africa; mackerel, herring and Dover sole from the UK; halibut, sablefish and toothfish from the US, and rock lobster (pictured) from Australia and the US. For more information, including where to buy, visit:

MSC www.msc.org

Marine Stewardship Council

with scores of different fish and fisheries. The site also offers a download-able wallet-sized guide to provide help in shops and restaurants. The information from this mini-guide has been reproduced below.

A third approach is to buy organic fish. These are, by default, farmed (nothing caught in the wild can qualify as organic, since you don't know where it has been or what it has eaten), but in much less intensive conditions, with no antibiotics or colorants and better protection for the environment. Devotees also swear that organic fish taste much better, as their diet is superior and the lower stocking density allows the fish to exercise properly. If your local shops don't sell organic fish, try the organic delivery companies listed on p.263. Bear in mind, though, that even when farmed organically, salmon, trout and other carnivorous fish are nonethe-less fed on fishmeal from sea-caught fish, so you're not really sidestepping the problem of the oceans' ever-depleting fish stocks.

A good fish guide

The following table, reproduced by kind permission of the Marine Conservation Society, provides an at-a-glance buying guide to sustainable fish and seafood. The criteria are fairly strict – so much so that even some MSC-certified products don't make it into the column designating fish that can be eaten with a totally clear conscience.

In the table, italics indicate the better choices within each column. So an italic item in the clear-conscience column designates the most sustainable fish available; an italic item in the caution column indicates a slightly greener choice than a non-italic item in that column.

Regardless of what species you buy, try to avoid young, undersized fish, as catching these can exacerbate pressure on stocks.

	Eat with a clear conscience	Caution: don't eat too often	Avoid
Abalone	Farmed	–	–
Alaska or walleye pollock	MSC certified	–	–
Anchovy	–	Portuguese coast	Bay of Biscay
Bib or pouting	All	–	–
Black bream, porgy or seabream	All	–	–

	Eat with a clear conscience	*Caution: don't eat too often*	*Avoid*
Clam	*American hardshell and manila, hand-gathered farmed sources*	*Dredge caught*	–
Cockle	MSC certified or hand gathered	*Dredge caught*	–
Cod, Atlantic	–	*Wild caught from N.E. Arctic;* Wild caught from Iceland, West English Channel, Bristol Channel, S.E. Ireland & Sole	Wild caught from all other areas
Cod, Pacific	*MSC certified*	–	–
Coley or saithe	*MSC certified*	*From Iceland or Faroes*	–
Crab	*Edible/brown, pot caught from South Devon;* Spider, pot caught	*Edible/brown, pot caught*	–
Dab	All	–	–
Dublin Bay prawn, scampi or langoustine	*MSC certified*	*W. Scotland, N. Sea, Skaggerak & Kattegat*	From Spain & Portugal
Flounder	All	–	–
Gurnard	Grey & red	*Yellow or tub*	–
Haddock	–	*N.E. Arctic & N. Sea, Skaggerak & Kattegat*	From Faroes and W. Scotland
Hake	–	*Cape, MSC certified;* European, northern stock	European, southern stock
Halibut	–	*Pacific, MSC certified*	Atlantic, wild caught
Halibut, Greenland	–	From N.E. Arctic	From N.W. Atlantic & Greenland, Iceland, W. Scotland & Azores
Herring or sild	From Norwegian stocks	*MSC certified*	South Clyde, W. Ireland & Great Sole fisheries

	Eat with a clear conscience	Caution: don't eat too often	Avoid
Ling	–	Handline caught from Faroes	All other stocks
Lobster	*Western Australian rock, MSC certified*	*European, pot caught*; European, net caught	From Canadian & S. New England stocks
Mackerel	MSC cert; handline-caught mackerel from the North East Atlantic	–	–
Monkfish or anglerfish	–	*From S.W. UK & N.E. USA*	N./N.W. Spain & Portugal
Mussel, common	*Rope grown or hand gathered*	–	–
Oyster	*Farmed native & Pacific*	–	–
Pollack or lythe	All	–	–
Plaice	–	*Otter trawled from Irish Sea, or gill/seine net from N. Sea*	W. English Channel, Celtic Sea, S.W. Ireland & W. Ireland
Prawn	Coldwater, from N.E. Arctic	Tiger, organically farmed; Coldwater, from N. Pacific	Tiger, farmed & wild caught
Ray	–	*Cuckoo and spotted from North Sea, Skaggerak, E. English Channel & Celtic Sea; Starry from North Sea, Skaggerak, & E. English Channel*	Smalleyed & thornback from Bay of Biscay; all blonde, sandy, shagreen & undulate
Red mullet	From N.E. Atlantic	From Mediterranean	–
Salmon	*Pacific (5 species are MSC certified); organic*	Atlantic, farmed	Atlantic, wild caught
Sardine or pilchard	From Cornwall	From Spain & Portugal	–
Seabass	Line caught	Farmed	Pelagic trawled
Seabream	–	*Organic*	Red or blackspot
Skate	–	–	All

	Eat with a clear conscience	*Caution: don't eat too often*	*Avoid*
Snapper	*Red (from W. Australia trap fishery)*; Malabar blood from W. Australia	*Vermillion and lane*; Silk and yellowtail	Cubera, mutton and Northern red
Sole	Common or Dover, MSC certified; Lemon, otter trawl caught	*Common or Dover, from E. English Channel and S.W. Ireland*	Common or Dover from North Sea & Irish Sea
Squid	Jig caught	*Trawl caught*	–
Swordfish	–	–	Indian Ocean, Mediterranean, and Central & W. Pacific
Tilapia	Farmed	–	–
Trout	Brown or sea and rainbow, organically farmed	Brown or sea and rainbow, farmed	Brown or sea, wild caught from Baltic
Tuna, albacore	*MSC certified*	*From N. Pacific*; pole & line from N. Atlantic	Longline & pelagic trawled from Mediterranean, N. & S. Atlantic
Tuna, bigeye	–	Handline and pole & line from Central and W. Pacific	All other stocks
Tuna, bluefin	–	–	All
Tuna, skipjack	Pole & line from W./Central Pacific or Maldives	*From Indian Ocean*	Purse seine from W. Atlantic
Tuna, yellowfin	–	Purse seine from Indian Ocean or E. Pacific; *all other stocks*	–
Turbot	–	*Farmed*	Wild caught

Fruit & vegetables

Buying them & growing them

Everyone agrees that a healthy, environmentally friendly diet is rich in fruit and vegetables. But fruit and veg varies widely in its climate impact, depending on how it is grown and transported. So what should we buy? Two key issues – the pros and cons of organic farming, and the impact of food transport – were discussed in chapter thirteen. This chapter raises a few other relevant issues, such as pesticide residues and labour rights, and also provides some basic guidance for getting started with growing your own fruit and vegetables.

Buying fruit & vegetables
What to consider

For the reasons discussed in chapter thirteen, the basic rules of thumb for choosing low-carbon, eco-friendly fresh produce are as follows: favour local, non-exotic seasonal produce; avoid air-freighted foods but don't worry so much about fruit and vegetables transported by ship; and favour organic when possible.

Very often, following this advice equates to favouring inexpensive produce. Vegetables transported by plane, or grown in heated greenhouses, will typically be more costly to produce than those items produced without large volumes of fossil fuels. (The exception is the organic component; organic produce is always a bit more expensive.)

What's in season when?

Choosing local, seasonal produce is good for two reasons. It reduces the need for long-distance transport and – assuming you stick to traditional varieties suited to your local climate – it ensures that there will be no need for energy-hungry heated greenhouses. Local, seasonal foods tend to score highly for nutritional value, too.

Even for imported foods it makes sense to consider the seasons. For example, oranges are most widely available in Europe during the winter months. And though some tropical fruits – such as bananas – are available all year round, it arguably makes environmental sense to buy more of them at times when little local fruit is in season.

The chart on the opposite page provides an at-a-glance breakdown of what's in season throughout the UK year. Various websites provide more in-depth information. Eat the Seasons, for example, highlights those foods that are at their best in any given week, and recommends seasonal game and fish as well as fruit and vegetables.

Eat the Seasons www.eattheseasons.co.uk

If you find seasonal food somewhat uninspiring, a cookbook arranged by the seasonality of the main ingredients – such as *River Café Green* – might help. And if you'd rather not have to think about what's in season when you go shopping, consider signing up for an organic box scheme (see p.263) or visit your local farmers' market (see p.261). Either way, you'll usually be offered mostly seasonal foods.

Pesticide residues

With thousands of farm workers killed every year by poisoning, and wildlife and the environment suffering in numerous ways, pesticides unequivocally pose certain threats to people and planet. Indeed, as discussed in chapter thirteen, the impact of pesticides on farm workers in the developing world is one good reason to favour organic when purchasing imported fruit and vegetables.

But what about the impact of pesticide residues on the consumer? Green campaigners link these residues to skin and eye irritation, mental and nervous problems, breast cancer and other conditions, and point out that little is known about the long-term effects of many of these substances (some of which are hormone disrupters), especially their combined "cocktail effect".

Seasonal fruit and vegetables in UK/Ireland

Adapted from BestInSeason.ie and other sources

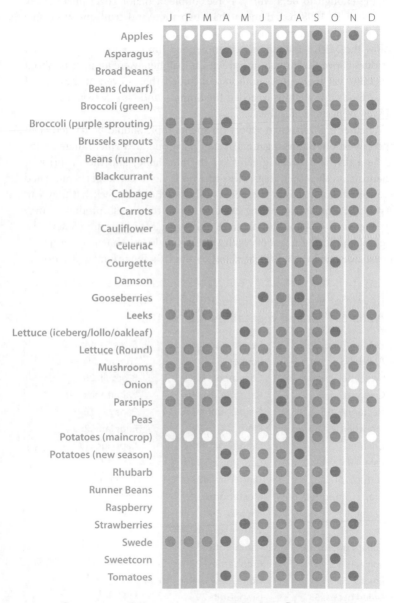

● = in season ● = coming into or out of season ○ = from storage

Most toxicologists seem less convinced that the quantities in question are big enough to be a worry. For example, a major government study published in 1998 tested more than two thousand fruit and vegetable samples and found that around 73% were residue-free, 26% had residues below the MRL ("maximum recommended limit") and 1.3% contained residues above the limit. Some green campaigners were up in arms about the 1.3%, yet the report's authors concluded that – even in the case of those foods that slightly exceeded the limits – "food is safe from the point of view of pesticide residues".

Whether a government report on such a topic should be trusted is an open question, but as a general rule this is one area where there seems to be a divide between greens, on the one hand, and the majority of scientists on the other. Regardless of who is right, consumers concerned about pesticide residues – especially those who don't feel that they can justify the expense of buying organic for all their fresh produce – may be interested to know which fruit and vegetables typically contain the highest and lowest levels of residues. The box below shows the results of a pesticide residue study carried out by the US's Environmental Working

Levels of pesticide residues found in various types of fruit and vegetables (out of 100)

Peach 100	Summer squash 53	Sweet potato 29
Apple 93	Pepper 51	Tomato 29
Sweet bell pepper 83	Cucumber 50	Broccoli 28
Celery 82	Raspberries 46	Watermelon 26
Nectarine 81	Grapes, domestic 44	Papaya 20
Strawberries 80	Plum 44	Eggplant 20
Cherries 73	Orange 44	Cabbage 17
Kale 69	Cauliflower 39	Kiwi 13
Lettuce 67	Tangerine 37	Sweet peas, frozen 10
Grapes, imported 66	Mushrooms 36	Asparagus 10
Carrot 63	Banana 34	Mango 9
Pear 63	Winter squash 34	Pineapple 7
Collard greens 60	Cantaloupe 33	Sweet corn, frozen 2
Spinach 58	Cranberries 33	Avocado 1
Potato 56	Honeydew melon 30	Onion 1
Green beans 53	Grapefruit 29	

Group (EWG). Note that the figure for each fruit or vegetable – measured relatively, out of one hundred – describes the residue levels on food as typically prepared, *not* the amount used in growing. So banana, for example, because of its thick, inedible skin, comes on the bottom half of the list despite being associated with intensive pesticide use. In other words, these figures relate to potential health risks to consumers – not to the total impact on the environment or farmers' health.

For more information about the EWG study, see their Food News website. Or to read more about the concerns that green groups have about pesticide use, visit Pesticide Action Network.

EWG Food News www.foodnews.org
Pesticide Action Network www.pan-uk.org

Fairtrade fruit and vegetables

Consumer-focused campaigns around worker rights have typically focused on sweatshops in the clothes sector, but similar issues have been associated with the production of fruit and vegetables. The most shocking labour abuse in this sector has occurred in the developing world, such as on the banana plantations of Latin America (see box overleaf). However, a number of exposés – such as those in Felicity Lawrence's book, *Not on the Label* – have shown that even European farms, including those in Britain, rely on temporary, non-resident workers subcontracted through gangmasters and paid sub-minimum-wage rates.

For fresh produce grown in the developing world, an increasing number of Fairtrade-certified fruits, and one or two vegetables, are available. These ensure decent terms and conditions for workers, just as with Fairtrade coffee and other certified products (see p.205). At the time of writing, the range of Fairtrade fresh produce includes apples, avocados, bananas, clementines, coconuts, grapes, lemons, lychees, mangoes, oranges, pears, pineapples, plums and satsumas. However, each item is only available from certain shops, according to market availability and seasonality. For more on Fairtrade fruit, see:

Fairtrade Foundation www.fairtrade.org.uk

For a brief while it looked as if the Fairtrade Mark was going to start appearing on certain British-grown organic foods. This was an attempt to deal with the fact that, even in comparatively wealthy countries such as the UK, ecologically friendly growers sometimes struggle to cover the cost of production. In the event, the idea was dropped, because Fairtrade

Bananas: globalization in a slippery skin

The world's most widely consumed fruit provides a perfect illustration of many of the debates surrounding intensive farming and world trade. There are hundreds of varieties of banana grown around the world, most of them in India and other countries where they are cultivated for domestic consumption. But the export market, driven mainly by the US and Europe, is dominated by just one variety and a handful of giant companies. More than half of banana exports are controlled by just three companies – Chiquita, Dole Food and Del Monte – which produce primarily in Central and South America, the world of the so-called "dollar banana".

Banana farming was once dominated by the United Fruit Company, which controlled much of Central America and was instrumental in bringing about the 1954 CIA-orchestrated coup in Guatemala, which overthrew the country's elected government and led to half a century of bloodshed. Today, the big banana firms don't stand accused of starting "real" wars, but they were behind the trade war which saw the US, Ecuador and various other South American countries battling with the EU at the World Trade Organization in the late 1990s. The point in question was the legality of the EU's banana import policy, which to a certain extent shunned "dollar bananas" in favour of those grown in ex-European colonies in Africa, the Caribbean and the Pacific (the "ACP" countries).

Pro-free-trade commentators argued that Ecuador and other dollar-banana countries were suffering due to their restricted access to European markets. But the EU claimed that small ACP family farms – and in some cases whole countries' economies – would be unable to survive without preferential treatment, so the human consequences of leaving everything to the free market would be huge. Anti-globalization protesters agreed, describing the situation as a classic case of big companies trying to use global trade rules to profit from a "race to the bottom". They claimed that the Latin American producers were only able to undercut small farmers in the ACP countries because their plantations depended on exploitation of workers and the environment.

It's true that the big banana producers have long been dogged by allegations about their low environmental and ethical standards. First, there's the question of wages: in the dollar-banana plantations, the workers who actually grow the bananas receive only 1–3% of the final price, or, put another way, as little as a dollar a day, in return for twelve or more hours of hard labour (according to figures from the Fairtrade Foundation). Child labour has been shown to be widespread, as has the intimidation, firing or even murder of would-be union organizers. Moreover, health and safety – and environmental sustainability – has often been seriously compromised by the use of agrochemicals.

Required in vast quantities on large banana plantations, fungicides and other pesticides are often sprayed from planes, and though there are rules to ensure that workers are not put in danger, these have not always been observed. As well as directly breathing in the chemicals, the workers, many of whom live on or next to the plantations, have often ended up drinking and bathing in contaminated

A banana worker from the Chinandega, Nicaragua, where widespread genetic illnesses, ranging from skin disorders to birth defects, are blamed on decades of pesticide use

Photo: Corbis

water. In the last few decades thousands of workers have died, been made seriously ill or given birth to deformed babies. According to a Fairtrade Foundation report from 2000, "some 20% of the male banana workers in Costa Rica were left sterile after handling toxic chemicals", while women working in the pack houses had twice the normal incidence of leukaemia. The ecological damage of the big plantations has also been severe, including soil degradation, lost biodiversity, toxic run-off and forest clearance. None of this was deemed relevant to the WTO case, however, since international trade rules don't take social and environmental issues into account.

Partly as a result of the "banana wars" (which the EU eventually lost) the Fairtrade labelling scheme was extended to cover bananas. Sourced from small-producer cooperatives in the ACP countries, fairly traded bananas provide all the standard guarantees of the Fairtrade Mark, including improved health and safety, basic environmental safeguards, and a higher proportion of the price going to the farmers. In response, the big banana firms gradually started trying to redefine themselves as socially responsible companies, publishing codes of conduct relating to health and safety, pesticide use and union representation. Chiquita even joined the UK's Ethical Trading Initiative and launched a partnership with the Rainforest Alliance. The initiatives are very welcome, though organic, Fairtrade bananas remain the only ones to provide any real green or ethical guarantees.

Bananas, it's worth adding, are usually shipped, rather than flown, from their country of origin to their point of sale. So the fact that they're grown thousands of miles from where we buy them doesn't add greatly to their carbon footprint. All told, an organic, Fairtrade banana is a relatively people- and planet-friendly fruit.

consumers tended to believe the scheme should prioritize the extreme poverty of the developing world.

In response, the Soil Association launched its own Ethical Trade label, with the aim of introducing a Fairtrade-style certification system – focusing on "fairness, mutual respect and transparency" – into the UK organic market. The key principles are: a fair price for the farmer and others in the supply chain; fair treatment of workers; and involvement in the local community. The scheme hasn't caught on in a big way, but you might see the label around from time to time.

Local vs organic vs Fairtrade

One common green conundrum is the question of whether it's better to buy organic produce or local produce. Very often it's possible to do both at once, but in some cases you have to pick between one or the other. Environmental groups tend to advise consumers to favour local over organic food, if it comes down to a choice between the two, but obviously it depends on how far the organic produce has travelled and whether it's been transported by ship, road or plane.

Sorting mangoes in a Fairtrade-certified plantation in Chacras, Ecuador

Fairtrade fresh produce can also occasionally present a dilemma. There's an increasing amount of fruit available that's organic *and* Fairtrade, but you sometimes have to plump for one or the other. In this case, there's a strong argument for favouring the Fairtrade option, since the certification process keeps a check on environmental sustainability (and indeed encourages growers to move towards organic certification) in addition to providing social and economic guarantees. But if pesticide use is firmly at the top of your priority list, then organic may be the way to go.

The Fairtrade versus local dilemma doesn't come up as frequently, since most fairly traded fresh produce consists of tropical fruits that couldn't be grown in the UK anyway. But since 2003, Fairtrade apples, shipped from South Africa, have been available, creating a slight tension between the Fairtrade movement, on the one hand, and environmentalists and UK farmers on the other. The apples from South Africa support empowerment projects, through which landless workers can become co-owners of fruit farms – a scheme praised by Nelson Mandela among others. Whether such positive schemes justify shipping the apples from the Southern Hemisphere at a time when many British orchards are being abandoned is for each consumer to decide.

Growing your own
How to get started

The ultimate in environmentally friendly food comes not from a farmers' market or box scheme but straight from a garden or allotment. Home-grown food cuts down on carbon emissions by reducing the need for synthetic fertilizers, pesticides, farm machinery, transport and packaging. As a bonus, it's also tasty and nutritious.

As food prices have risen and concern about climate change has mounted, an increasing number of people have rediscovered the pleasures and benefits of growing their own. For some people – such as those with minimal outdoor space – it's practical only to grow a very small proportion of their fruit and vegetable requirements. But with a little commitment and a decent patch of sunny garden it's perfectly possible to significantly reduce the amount of fresh produce you need to buy, thereby reducing your carbon footprint as well as your food bills.

A full guide to growing your own is beyond the scope of this book, but the following pages provide the key information you need to get started.

Pick your plot

What you can grow depends on two key things: the amount of time and effort you're prepared to put in; and the kind of space you have available, including how much sun it gets and what the soil is like.

Allotments

Ever since the late nineteenth century, British law has required local authorities to provide residents with low-rent plots – allotments – for growing fruit, vegetables and flowers. In the first half of the twentieth century, more than a million active allotments provided a significant slice of the UK's fresh produce. After World War II, numbers fell rapidly right through to the turn of the millennium, since when a strong revival of interest, fuelled partly by rising environmental awareness, has led to demand significantly outstripping supply in many areas.

Today, some councils have waiting lists running to many years, but in other places you might be able to secure an allotment in a matter of weeks or months. Even if you have to wait more than a year, it can be well worth it, as an allotment will give you enough space to grow a large proportion of your fruit and vegetables in return for a rent as low as £20 per year. Contact your local council for more information.

Gardens

If you have sufficient outside space, a decent kitchen garden can be even better than an allotment. Having your plants just outside the home makes cultivation and watering much easier – and helps reduce car use, too. Even in smaller gardens, you can get a surprising amount from a modest-sized vegetable patch or a couple of raised beds.

When deciding which bit of your garden to dedicate to vegetables, think first and foremost about sun. Many food plants – such as tomatoes – will thrive almost in proportion to the amount of sunlight they get. The best plots get sun for most of the day, though you can achieve a lot with a few hours of uninterrupted sun in the growing season. West-facing plots are better than east-facing plots, since plants receiving limited light tend to prefer it in the warm afternoon rather than the cool morning.

The next consideration is the soil. If your garden already has rich, fertile, moist but well-drained earth then you have a huge head start. If the ground is sandy or rocky, dry or waterlogged, then you might consider building a couple of raised beds and filling them with fresh compost.

Bed layout and crop rotation

There are many ways to lay out a fruit and vegetable growing area, and the best solution for each garden will vary according to light and other factors. If you have adequate sunny space, however, a sensible strategy is to create multiple beds, each a couple of metres square and surrounded by walkways big enough to let a wheelbarrow through. The walkways reduce the chance of trampling on crops and save you needlessly getting muddy, while the separate beds allow you to practise crop rotation – a key concept in home-grown food just as it is on organic farms. The aim of crop rotation is to alternate the types of plants grown in any particular patch of soil each year. This stops the build-up of pests and avoids depleting the fertility of the soil.

Different gardeners favour different patterns for their crop rotations. Some rotate five or even six different categories of vegetables between the same number of beds. For most gardeners, though, a three-way crop rotation is sufficient. In this arrangement, each bed would rotate these three crop families:

Legumes & onion family Peas, beans, garlic, leeks, onions, shallots, garlic

Roots Potatoes, carrots, celery, parsnips

Brassicas Broccoli, cabbage, cauliflower, radish, swede

If you have sufficient space, the ideal arrangement is to have three separate beds; this way you get to have all three of the above categories growing every year, as shown below. Other vegetables, such as tomatoes and squashes, can be grown whenever there is space, without worrying about rotation.

	Bed 1	**Bed 2**	**Bed 3**
Year 1	Legumes/onions	Roots	Brassicas
Year 2	Brassicas	Legumes/onions	Roots
Year 3	Roots	Brassicas	Legumes/onions

Small gardens, patios, balconies and windowsills

If your space is limited to a patio, balcony or even a windowsill, it's still perfectly possible to grow some fruit and vegetables. Even a humble windowbox can provide a surprisingly prolific crop of herbs and cherry tomatoes. (The tomatoes shown on the back cover of this book were grown by the author on the windowsill of a second-floor London flat.) If you have a patio with space for larger pots then you may get good results with peppers, chillies, aubergines, courgettes, beetroot, carrots and potatoes. Salads are well suited to containers of all sizes – particularly the "cut and come again" varieties that allow you to harvest just a few leaves at a time. As for fruit, strawberries can thrive in pots, and so can some fruit trees, including peaches, apricots and nectarines.

Most of these plants will do perfectly well with standard multi-purpose compost, though it's a good idea to add water-retaining crystals (or to buy a compost that comes with crystals). If you're using terracotta pots, water retention can be further improved by lining each pot with polythene, though be sure to make some holes at the bottom to allow sufficient drainage.

What to grow

The following pages provide some basic background and advice on growing a selection of the most popular fruits and vegetables, drawn from this book's sister volume, *The Rough Guide to Food*.

Apples

Plant November–February (dwarf varieties in containers)

Harvest July–January

Likes Compatible trees for pollination; feeding in spring, mulch in April

Dislikes Waterlogged soil, cramped branches (prune to thin these out); pests: codling moths, sawfly, earwigs and wasps

Varieties James Grieve, Coxes Orange Pippin, Egremont Russet

Tips Dig a very big hole, cover with mulch and keep new trees the recommended distance apart. Remove all blossom the first year

Asparagus

Plant March–April. Plant ten one-year-old crowns 15–20cm deep, 30cm apart in a single well-prepared row

Harvest May–June

Likes Very well-drained and slightly alkaline soil, clear of weeds, seaweed fertilizer

Dislikes Cold and wet conditions, slugs

Varieties Connover's Colossal, Backlim, Gijnlim

Tips Although slow to start, once established these perennial plants are trouble-free and last for up to twenty years. Parsley can be grown between crowns

Broad beans

Sow Autumn: sow seeds inside in October, then plant out small plants (10cm high) November–December. Spring: sow seeds directly into the soil in February–April, about 20cm apart. Final height: 1m

Harvest May–September

Likes Fertile soil, cool conditions

Dislikes Blackfly. Pinch out the plant's tips (which pests like) when in full

Organic blueberries on the bush

flower, brush off or spray with soap mix

Varieties Aquadulce or dwarf variety "The Sutton"

Tips Leave a few weeks between the planting of rows to enjoy tender small beans through summer. Brown leaf marks indicate a soil potassium deficiency but won't harm the crop

Blueberries

Sow November–February in 40cm diameter container

Harvest Three years after planting, July–September

Likes Sunny, sheltered spot, acid soil (pH5 4–5.5), mulch with bark

Dislikes Tapwater (use rainwater), birds (use nets)

Varieties Duke (excellent all-rounder), Colville, Brigitta (late fruit), Toro (for small spaces)

Tip Grow three varieties alongside each other for optimal cross-pollination. Fairly trouble free

Broccoli

Sow March–May (under cover for first month). Plant in 2–3 batches of 10–20 small plants, about 20–40cm apart (depends on variety) in rows

Harvest June–October (Calabrese); January–April (purple sprouting)

Likes Nitrogen-rich soil, sheltered site, lots of space (purple sprouting)

Dislikes Slugs, snails, white fly – might need protecting from pests with netting

Varieties Early Purple Sprouting Improved, Belstar (Calabrese)

Tips Can suffer from disease "clubfoot", which remains in soil for up to twenty years

Courgettes

Sow Mid-April or May (under cover) and June into soil. Plant seeds, thinnest edge facing up/down (or small plants) into 30cm compost-filled holes

Harvest July–October, ten weeks after sowing

Likes Lots of water and space (about a square-metre per plant)

Dislikes Frost, red spider mite and mosaic virus

Varieties Green Bush, El Greco, Gold Rush (yellow fruits), Little Gem (small round fruits on a trailing plant)

Tips Expect around fifteen fruits per plant. Stagger the sowing to avoid a glut or find a good chutney recipe. Use baby courgettes along with the yellow flowers (and onion and garlic) to make a delicate risotto; left too long courgettes swell to become marrows

Endive

Sow February–September

Harvest All year, 7–13 weeks after sowing

Likes Moist soil

Dislikes Slugs

Varieties Curled or frisée: Pancalieri, Monaco. Broad-leaved: Scarola

Tips Easier to grow than chicory

French beans

Sow Sow in 2–3 batches (for regular picking) from April (under cover) to August

Harvest When they are the diameter of a pencil in about July–August

Likes Sunny spot and moist, rich soil. Support climbers with a wigwam structure

Dislikes Slugs, blackfly, frost

Varieties Dwarf (bush): Delinel, Royal Burgundy (purple). Climbing: Blue Lake

Tips Very productive and easy to grow, dwarf and climbing varieties are good for small spaces and they freeze well. Try pink and white striped Borlotti beans, too

Leeks

Sow Grow small (15cm) plants from seed indoors, ready for planting out into deep, narrow holes in June; sow seeds outside in May with 10cm between plants.

Harvest September–May

Likes Sun, water

Varieties Musselburgh (very hardy, dates back to 1834), Apollo

Tips "Earth up" the emerging leek with soil to increase amount of white stem. Closer spacing of plants produces smaller leeks

Lettuce

Sow Germinate seeds indoors for planting out February–September

Harvest All year

Likes Very rich soils with lots of nitrogen

Varieties Cos: Corsair, Little Gem (takes over two months to develop a heart). Loose-leaved: Lollo Rosso, Lamb's lettuce (small rounded green leaves), Catalogna (serrated leaves), Salad Bowl (green and dark red oak leaves)

Mizuna

Sow March–August (outside); September–February (under cover)

Harvest All year, 3–8 weeks after sowing

Likes Moist soil. Keep well watered

Dislikes Dry conditions; will rot if not protected against heavy rain and snow

Varieties Tokyo Beau and Tokyo Belle

Tips Sow some to harvest as whole plants (25cm apart) and some to cut-and-come-again (10cm apart, you will

get about five pickings). Can be grown through winter

Potatoes

Sow Earlies March–May; maincrop mid to late April. Plant chitted potatoes 40cm apart into a 25cm deep, 30cm wide trench with dug-in manure at its base

Harvest When flowers die down, the potatoes are ready to pick. Earlies: June–July. Maincrop: September (store over winter in a cool, dry environment)

Likes Rich soil and potash (wood ash from stove or potash-rich comfrey); regular watering while flowering

Dislikes Shade, frost, blight fungus

Varieties First earlies: International Kidney (aka Jersey Royals). Second earlies: Kestrel or Wilja (high yield and good disease resistance); Charlotte (classic French salad potato). Maincrop: Belle de Fontenay (old French early

Digging up potatoes

maincrop); Santé (excellent for organic growers as very resistant to pests and disease)

Tips "Earthing up" the plants stops the tubers being ruined by sunlight and turning green. To do this, cover the plants with soil until only 10cm or so of foliage is visible. Expect to do this a few times each growing season

Rocket

Sow All year round indoors; in summer outside

Harvest All year, 3–4 weeks after sowing

Dislikes Bolts (flowers) in hot weather

Varieties Avanti

Tips Very easy to grow; minimal pest problems

Spinach

Sow Regular small sowings every three weeks from March–September

Harvest April–November

Likes Cool conditions

Varieties Galaxy, Bloomsdale

Tips Can bolt (flower) in hot weather. If you're having problems try chard (also known as "perpetual spinach") instead

Tomatoes

Sow Seeds (March–April) for planting out as small hardened-off plants (May/June)

Harvest July–October

Likes Fertile, well-drained soil, plant food and sun (a south-facing wall is perfect), regular light watering (to prevent splitting of skin)

Dislikes Shade, white fly (plant marigolds to deter them)

Varieties Alicante, Green Zebra (green/yellow stripes). Cherry: Cherry Belle, Sungold (yellow), Gardener's Delight. Hanging baskets: Tumbler or Pearl; mix historic varieties

Tips Keep plants focused on fruit (not leaf) production: pinch off non-flowering sideshoots and the top of plant when fruiting. Green tomatoes make excellent chutney

Green gardening: edibles and beyond

With so much agricultural land being given over to monocrop industrial farming, some commentators hope that gardeners could be the saviours of biodiversity in densely populated countries such as Great Britain. Cultivating a garden or allotment with a wide range of edible and non-edible plants and trees can certainly provide sanctuary for a wide range of insects, birds and other creatures.

On the other hand, certain garden products can have the opposite effect, causing harm to the environment either through their toxic contents or through their extraction from the earth.

Pesticides and alternatives

As the abundant health warnings on the packets attest, many garden herbicides and insecticides contain fairly nasty substances. Some of these not only hit the target species but also poison birds and other wildlife and pollute local water systems. As with many household chemicals, they're also likely to be the result of a polluting manufacturing process.

Just as with farming, however, weed and pest killers can be largely avoided with a bit of effort. For weeds, this generally means pulling or digging them out by hand, but if an area has been invaded by pernicious weeds (such as bindweed or ground elder), the best bet is to dig over the ground and then cover it with old carpet or plastic in order to deny the plants light. If the weeds haven't taken over completely, a less drastic measure is to cover the dug ground with a layer of newspaper and then a light-excluding mulch such as wood chippings.

Pests – in particular slugs and snails – can be the bane of the gardener's life. Fortunately the natural solution is often the best. Birds (like thrushes) will crack open and eat snails, slugs are part of the hedgehog's diet, while ladybirds and lacewings keep aphids in check. You can encourage birds and hedgehogs either by the way you plant (birds will nest in thick hedges) or by having strategically placed nesting boxes. If you must use slug pellets, make sure to buy a brand that is harmless to all other creatures.

Companion planting – the growing of specific plant species in close proximity – provides another option. For example, nasturtium can attract caterpillars away from adjacent cabbages, carrots and leeks can repel each other's pests, and marigolds next to tomatoes can stop aphids attacking your crop. For more advice, see:

Companion Planting www.companionplanting.net

Compost & peat

Of the various garden products collected at cost to the environment, the worst is perhaps peat, which forms the basis of much of the compost on sale in garden centres and elsewhere. The problem is that harvesting peat means draining and digging up irreplaceable peatland bogs, which are home to a wide range of increasingly rare plant and animal species (and which also serve a valuable role in stabilizing ground-water levels). According to a 2003 report from two environmental groups – the WWF and Traffic – around two thousand hectares of peatland bogs are dug up in Ireland each year, mainly to supply British gardeners.

The advice of most green groups is to only buy compost specifically labelled as peat-free. Such composts make use of sustainable alternatives such as coir peat, a biodegradable byproduct of the coconut industry. Peat-free compost isn't difficult to find in the big home-improvement chains but it can be trickier in supermarkets and some independent garden centres. It's worth noting that a peat-free growing medium does not hold moisture as well as a peat-based one, so in summer you should water less heavily but more often.

Of course, the ideal solution is to make your own compost. This not only gets around the water issue but also helps minimize the amount of food you throw away. For more on composting see p.132.

More green gardening tips

▶ To minimize the use of a hose (and maximize your supply of plant-friendly rainwater) fit your garden with rain butts. As well as taking the water from the roof guttering of your house, these can be set up against garden sheds or outhouses. For more on home and garden water use, see p.135.

▶ Wooden garden furniture is usually made of hardwood, so be sure to look out for the FSC logo when buying; where the label isn't shown, at a minimum avoid teak and other tropical and subtropical hardwoods. See p.273 for more information about sustainable wood.

▶ Avoid "water-worn", "Irish" or "weathered" limestone. Limestone pavements (outcrops of rocks that have been weathered into a paving-stone pattern) support rare plants and wildlife, but most of those in the UK have been carved up and made into rockeries and water features.

▶ For seeds and other eco-friendly garden supplies, try one of the following sites. As its name implies, the Organic Catalogue is also available in paper form. Call 0845 130 1304 to order a copy.

HDRA www.hdra.org.uk
Green Gardener www.greengardener.co.uk
Organic Catalogue www.organiccatalog.com
Wiggly Wigglers www.wigglywigglers.co.uk

Where to buy food

Supermarkets & alternatives

The explosive growth of supermarkets in the last half-century has completely revolutionized the food sector. If Britain ever was a nation of shopkeepers, as Napoleon supposedly quipped, that era has long-since passed. Today, just four companies – Tesco, Sainsbury's, ASDA and Morrisons – sell more than three quarters of our food. Along with a handful of smaller players, including Marks & Spencer, Waitrose, the Co-op, Iceland and Budgens, the supermarkets – or "multiples" as they're known in the trade – account for the overwhelming majority of grocery sales.

Despite the supermarket stranglehold, some markets and local shops survive. And a number of alternatives with a distinctly green or ethical slant – such as farmers' markets, organic box schemes and online specialists in Fairtrade products, free-range meat and so on – are gaining popularity across the country. Following is a brief look at the green debates surrounding supermarket shopping, followed by a small directory of alternative outlets.

Supermarkets
Do their green claims stack up?

Perhaps more than any country in the world, the UK is addicted to supermarkets. Maybe it's their special offers, their dazzling choice or their all-under-one-roof convenience. Whatever the reasons, few Britons buy much of their food anywhere else. Of every pound spent on food in the UK, more than 85 pence goes into a supermarket till. Despite this enormous success, supermarkets are regularly criticized in relation to a wide range of environmental – as well as broader ethical – concerns. So how green are supermarkets compared with the alternatives? Should we believe their numerous environmental claims, or do they represent little more than greenwash?

The short answer is that although there are environmental problems with supermarket shopping, the important ones from a climate change perspective aren't those that get the most publicity. For example, despite huge media coverage, the giving out of plastic bags at supermarket tills is not a major concern in terms of global warming (see p.118 for more on this topic). The same is true for food packaging. There's no doubt that much food is grossly over-packaged in supermarkets (and indeed elsewhere), and it's true that all packaging causes some emissions in its manufacture, transport and disposal. Compared to the food it contains, however, most packaging isn't particularly significant in terms of climate change – especially when it's made of light materials, as in the case of foam trays and plastic wrap. The carbon footprint of a glass wine bottle – which is heavy and bulky to transport and energy-intensive to produce and recycle – is likely to be hundreds of times higher than that of the polythene wrap around a cucumber or bunch of grapes.

Moreover, it's worth bearing in mind that some packaging of fresh produce can help reduce the amount of food that's spoiled in transit and storage. Packaging can also increase the lifespan of the food, thereby reducing the chance that it will be thrown away in the home. In some cases, then, plastic packaging, however aesthetically unpleasant it might be, may actually be beneficial in terms of climate change, since the emissions generated in its production will be lower than that of the wasted food it avoids. Unfortunately, it's not always easy or possible to know which packaging is wasteful and which is serving a useful purpose. But the fact remains that, while there may be other reasons to object to it, light plastic packaging isn't likely to be a key issue in terms of the carbon footprint of your diet.

Another popular criticism of supermarkets is that they sell products high in "food miles" (see p.193). It's probably true that supermarkets are worse than other outlets in this regard. In particular, it seems likely that the supermarket retail model, with its fast turnover and focus on premium products, has done the most to drive the increasing number of air-freighted products being consumed in the UK. To be fair, however, food transported by air can be found in smaller shops and indeed markets,

Banks of doorless fridges and freezers are one reason why supermarket stores consume so much energy

too, so the big retailers can't be held entirely responsible. Furthermore, as we've seen, air-freight accounts for only a tiny proportion of the food and groceries we buy. When it comes to the ships and lorries that deliver the rest, no one knows with any certainty how supermarkets compare with other outlets. About all that it's possible to say for sure is that food transport emissions will usually be lower if you get your fresh produce via a farmers' market (presuming you don't need to drive too far to get there) or an organic box scheme that consciously keeps food miles to a minimum.

One less discussed but probably fairly significant issue is refrigeration. Supermarkets sell a large and growing proportion of their food in the form of pre-prepared readymeals. Although there isn't a great deal of data available to prove the point, it seems highly likely that buying ingredients and cooking them at home will typically cause fewer emissions than buying a readymeal that has been created in a factory, transported in a chilled lorry and then displayed in an open-fronted fridge or freezer.

Doorless fridges and freezers aren't the only example of energy-profligacy in supermarket stores. As George Monbiot documented in his book *Heat*, the stores are also lit to unusually high levels of brightness and are often wasteful with heat, as in the case of warm air machines in entrance-ways. According to a study by ENDS (Environmental Data Services) and the University of Edinburgh, the operation of supermarket stores and transport fleets is directly responsible for around 1% of UK greenhouse gas emissions, even before you consider the food they sell.

A final environmental argument against supermarkets is their tendency to refuse to accept fruit and vegetables on pernickety cosmetic grounds. Size is specified to within a few millimetres, and perfectly natural bumpy or uneven shapes, and varying colours, are disallowed. Critics claim that this obsession with how produce looks leads to a great deal of waste, and encourages farmers to increase their use of agrochemical inputs.

All these arguments aside, it's worth bearing in mind that where we shop will typically be less important than what we buy. The carbon foot-print of a basket of seasonal vegetables picked up at the local supermarket will typically be significantly lower than that of a basket of organic, artisan meats and cheese from the local farmers' market.

Which supermarkets are the greenest?

As consumers have become more concerned about environmental issues, the major supermarkets have responded with a wide range of measures designed to win green hearts and minds. The race started a few years ago, when, in the space of a few months, Tesco started offering Clubcard points to those who bring their own bags; ASDA initiated a trial scheme whereby farmers deliver produce directly to their nearest branches, reducing food miles; Waitrose introduced dedicated local food sections; Sainsbury's announced a plant to cut food miles by 5%; and M&S unveiled a hugely ambitious scheme ("Plan A") that set five-year targets for emissions, waste, resource use, fair trade and healthy eating.

With such varied and fast-developing initiatives, it's not easy to rank the supermarkets in terms of their overall environmental standards. One comparison was carried out by the National Consumer Council in 2007. This assessment rated the various supermarket chains for performance on climate change, waste, sustainable fishing and sustainable farming, awarding each an overall grade between A (the greenest) and E (the least green). The scores were as follows:

▶ **M&S, Sainsbury's, Waitrose** scored B
▶ **Asda, Tesco** scored C
▶ **Co-op, Morrisons, Somerfield** scored D

Supermarkets and animal welfare

When it comes to animal welfare, the general trend is simple: the higher-end the supermarket, the better the policies. Hence in one recent survey the RSPCA congratulated "Marks & Spencer, Selfridges and Harvey Nichols" as the only supermarkets to ban battery eggs completely. The most comprehensive research into supermarkets and animal welfare is published annually by Compassion in World Farming, which compares the big retailers on a wide range of welfare criteria. The 2007 report ranked the main supermarkets as follows, out of a possible five points:

1. **Marks & Spencer** 3.99

2. **Waitrose** 3.98

3. **The Co-operative Group** 2.88

4. **Sainsbury's** 2.67

5. **Tesco** 2.63

6. **Morrisons** 2.36

7. **Somerfield** 1.89

8. **Asda** 1.86

(It's worth noting that if the supermarkets were compared on broader ethical factors – such as Fairtrade, household chemicals, animal welfare and animal testing – the Co-op would likely emerge as a high performer.)

Another study, by academics at the University of Edinburgh, published in mid-2009, compared the supermarkets specifically in relation to their greenhouse gas reduction targets. Perhaps surprisingly, given its reputation for pursuing profits at any cost, Tesco came top of the list. This reflects Tesco's plan to reduce emissions by 5.6% each year until 2020 – an ambitious target for a business of any size. The study acknowledged that M&S could yet outperform Tesco on emissions cuts, depending on progress with its carbon neutrality plan.

Alternative food retailers
Farmers' markets and organic delivery

Though they still only account for a fraction of a percent of UK food sales, farmers' markets are on the up. In the words of FARMA, the National Farmers' Retail and Markets Association, a farmers' market is a place where "farmers, growers or producers from a defined local area are present in person to sell their own produce, direct to the public. All products sold should have been grown, reared, caught, brewed, pickled, baked, smoked or processed by the stallholder." This approach aims to give the farmers a better deal than they would get via a retailer, to minimize environmentally harmful food transport, and to give consumers fresher produce and the chance to put questions directly to the producers.

To ensure these aims are kept pure, FARMA has developed a set of standards which farmers' markets must adhere to in order to gain official certification:

▶ **Local produce** "Primary produce" sold must be grown, reared or caught by the stallholder within the defined local area: usually a thirty mile radius around the market, though up to fifty miles is acceptable for larger cities and coastal or remote towns and villages, and in some cases the county is considered the definition of the local area. "Secondary" (prepared) produce must be brewed, pickled, baked, smoked or processed by the stallholder using at least one ingredient of origin from within the defined local area.

Reconnecting field and city: a farmers' market in central London

▶ **The principal producer** or a representative directly involved in the production process must attend the stall.

▶ **Information** should be available to customers about farmers' markets standards and the production methods of the producers.

Farm shops apply a similar philosophy on a smaller scale. Some offer the produce of just one farm; others include a full range of foods from local producers, presented in a converted barn or other farm building.

For a list of your nearest farmers' markets and farm shops, and for more information about how they work, visit:

Farmers' Markets www.farmersmarkets.net
London Farmers' Markets www.lfm.org.uk
FARMA www.farma.org.uk

Organic box schemes and delivery companies

A growing number of environmentally concerned consumers are choosing to have organic food dropped directly at their door. Typically the main purchase is a weekly box or bag of local, seasonal fruit and veg, bolstered where necessary by produce imported by ship. However, most delivery companies also offer all sorts of other foods and goods, from bread to washing-up liquid, so it's perfectly possible to get everything you need in one weekly drop and to avoid ever going to the shops.

Even when you factor in the delivery van, the overall footprint of an organic delivery is likely to be smaller than that of a supermarket shop – assuming, that is, that you stick to a local, seasonal selection of fruit and vegetables. And the more people that sign up, the bigger the benefits will be, because the delivery vans will be able to do more drops on each trip.

The Soil Association is the best source of information about box schemes and organic delivery companies. Their site lists hundreds within the Buy Organic section.

Soil Association www.soilassociation.org

If you can't find any local schemes, or if those you've tried aren't very good, you could try one of the bigger box scheme companies, some of which will deliver to nearly anywhere in the UK. For example:

Fresh Food www.freshfood.co.uk
Simply Organic www.simplyorganic.net

The Southeast is also well served by Organic Delivery (London only) and the award-winning Abel & Cole.

Organic Delivery www.organicdelivery.co.uk
Abel & Cole www.abel-cole.co.uk

A typical mixed box from Abel & Cole. Like most organic delivery services, this company offers a wide range of fruit and veg options as well as bread, fish, meat and store-cupboard goods.

Fair trade specialists

The Fairtrade Foundation's website – www.fairtrade.org.uk – lists all the certified Fairtrade products available in the UK, along with details of which shops stock them. However, for the best selection at the lowest prices, try an online specialist such as the following.

Traidcraft www.traidcraftshop.co.uk

Traidcraft sell a good range of fairly traded food, from basics (sugar, rice, pasta, nuts, jam, honey), via wine, teas and coffees, to a mouthwateringly rich list of chocolates and other sweet things. Many items are available in packs of six or more, allowing you to make savings when buying in bulk.

Goodness Direct www.goodnessdirect.co.uk

Goodness Direct aims to sell "everything for your healthy eating, natural and ethical lifestyle". That includes a range of more than 150 Fairtrade-certified products in addition to various veggie, vegan and health-food offerings and pet food. There is even a convenience food section, where the time-strapped can stock up on Organic Tofu Smoked Almond & Sesame Slices and the like.

Clipper Store www.clipper-teas.com

Probably the best way to buy Fairtrade-certified hot drinks is direct from Clipper, who offer an enormous range of teas as well as a decent number of coffees and hot chocolates. Most of what's available is also organic.

Other sites and stores

Following are a few other sites and organizations that sell food relevant to the issues discussed in this chapter.

Big Barn www.bigbarn.co.uk

This site aims to "help rebuild local food supply chains across the UK". Enter your postcode to be presented with a map showing your local food producers, specialist markets and other points of interest. The site also has a local food blog, a seasonality guide and a recipe finder to help you work out what to actually do with your local produce.

UK Food Online www.ukfoodonline.co.uk

Though geared towards high-quality food, rather than green and ethical issues, this site includes plenty of relevant suppliers. Also, since they don't limit themselves to organic producers, they include many suppliers not found in the Soil Association directories.

Suma Wholefoods www.suma.co.uk

Though they primarily supply to shops, this large wholesaler of green, organic and fairly traded foods (as well as other products) will deliver directly to customers who don't have access to a local retailer, subject to minimum orders.

Part V

Shopping
& services

Greener shopping
Money matters
Carbon offsetting

Greener shopping

Gauging the environmental impact of your purchases

So far, this book has focused primarily on home energy, travel and food. These are all key areas for anyone interested in living a greener life but, as we saw in chapter two, many additional categories of purchases and activities contribute to our carbon footprints. Indeed, virtually every pound we spend generates greenhouse gas emissions and some items – products containing toxic chemicals, for example, or made of timber logged from virgin forest – have an environmental impact over and above their contribution to global warming. There isn't space in this book for a comprehensive guide to green shopping, but this chapter covers a few key areas, starting with some general advice on how to minimize the carbon costs of your purchases.

Embedded carbon
Tips for low-carbon shopping

When we buy home energy, car fuel or plane tickets, it's fairly easy to work out the carbon footprint incurred. For other items, it's not so simple. As described in chapter two, it's immensely complicated to work out the precise carbon cost of even a simple product such as a bag of sugar; inevitably it's trickier still to work out the emissions involved in producing an item with a long and complex supply chain – a new kitchen, say, or a television. In some cases, you might find a carbon footprint label, but in other cases you'll need to either use intelligent guesswork or do some online research to get a rough sense of the "embedded carbon" or "embedded energy" in each item.

Carbon labels

Though they currently appear on only a small number of products, labels specifying the carbon footprint of a product are likely to become increasingly common in coming years. The standard carbon labelling scheme in the UK was developed by the Carbon Trust, a quango set up by the government to help businesses cut emissions. The label shows the total carbon footprint of a

working with the Carbon Trust

2.8kg

CO2

per garment

We have committed to reduce this carbon footprint

The carbon footprint of this product is 2.8kg. This is the total carbon dioxide (CO2) and other greenhouse gases emitted from the raw materials, production and transport to the UK

This compares to the carbon footprint of an identical product manufactured without the use of renewable electricity which is 28kg per garment

product in grams or kilograms of carbon equivalent (ie all the greenhouse gases emitted during the production of the item, expressed as CO_2 for the sake of simplicity).

In some cases extra information is provided on the label to help consumers get a sense of how the product compares to the alternatives. For example, the carbon labels that appear on the various types of Tesco orange juice state that freshly squeezed juice creates more emissions than juice made from concentrate. This isn't surprising, given that making fresh juice involves transporting the whole fruit to the processing plant, rather than just the juice, but the carbon label usefully quantifies the difference – 360 grams per serving, compared to 240 grams – and in doing so helps the consumer make an informed choice.

Some experts have expressed niggles about the methodology behind the Carbon Trust's labelling scheme but in most cases these are caveats rather than anything fundamental. One example is that the rules, which have been formalized by the British Standards Institute under the name PAS 2000, don't apply a multiplier to CO_2 emissions from planes on the grounds that the science isn't yet sufficiently clear. (For more information on aviation emissions and multipliers, see p.170.) This means that airfreighted goods look greener than they probably should. Overall, though, the rules are fairly robust and the labels are a useful way to raise carbon literacy among consumers.

Looking up carbon footprints

For the vast majority of items that *don't* feature a carbon label, it can be difficult to get even an approximate sense of their carbon footprints. Individual reports, articles and blog posts have focused on the emissions caused by the manufacture of various specific items, so sometimes it's possible to track down useful information via Google. Surprisingly, however, no single site pulls together footprint estimates for a wide range of objects. This can make it challenging not only to weigh up the carbon impact of different purchases but also to assess the environmental benefits of upgrading from old appliances to modern energy-efficient ones (a calculation that's impossible to make without knowing the embedded carbon in the new machines).

A reliable carbon footprint website will doubtless eventually emerge, but in the meantime the most useful starting place for researching the climate impact of objects is WattzOn. The site's Embedded Energy Database (EED) provides ballpark figures for the energy consumed in the manufacture of everything from dishwashers to footwear. Since energy use is the

Wattzon's Embedded Energy Database breaks down dozens of items into their constituent materials and gives figures for the energy that each takes to produce. The results are presented in a table similar to the ones used for nutritional information on food packaging. The energy figures are presented as joules – ie a constant flow of energy throughout the expected lifespan of the product.

main source of greenhouse-gas emissions for most manufactured goods, these "embedded energy" figures give a useful indication of the relative carbon footprints for the various items listed.

WattzOn EED www.wattzon.com/stuff

Another useful source for information about the carbon footprint of everyday objects will be Mike Berner's-Lee forthcoming book, *How Bad are Bananas? The Carbon Footprint of Everything*. Due to be published in 2010, the book details the climate impact of more than a hundred items.

Guesstimating carbon footprints

In most cases, when you're out shopping you'll neither see a carbon label nor have the time or inclination to research the embedded energy of the various objects you're thinking of purchasing. For that reason, it's useful to know a couple of rules of thumb for gauging the approximate carbon footprints of different types of object. One key factor is weight. This is important not only because it reflects the amount of material used to produce an object but also because the heavier an object is (and indeed the bigger it is), the more energy it will have taken to transport to the shop.

For some products, such as bottled water, weight is the only serious concern: items such as these use very little energy to produce, but a lot to transport. For most products, however, weight gives only part of the picture. Even more important are the materials used to produce the object.

Many green-minded individuals balk at plastic, because it's made from oil, includes various synthetic chemicals and is slow to degrade. From a climate change perspective, though, most plastics aren't particularly energy-intensive to make. Furthermore, plastic products tend to be relatively light. For both these reasons, the carbon footprints of plastic objects are usually fairly low.

By contrast, metals tend to be relatively carbon intensive. For one thing, it takes a huge amount of energy to mine and purify them – especially aluminium, which is separated from its bauxite ore using electrolysis, a process that consumes a massive amount of electricity. In addition, metal products tend to be heavy, which boosts their footprints even further.

At the other end of the spectrum is wood. If it's cut from well-managed forests, wood can even be "carbon positive", as it is largely made of carbon pulled out of the air. On the other hand, as we'll see below, wooden products made from timber that's been logged from virgin rainforest can be disastrous in terms of climate change and biodiversity.

Emissions per pound spent

Unless we reduce the amount that we work or give more to charity, each of us has a certain amount of money available to spend. Given this fact, the most important aspect of green shopping is arguably not the carbon footprint of each item that we buy but the carbon footprint of each pound we spend. Seen in this way, luxury items are greener than lower-quality items – not just because they're likely to last longer, but because they will typically have a lower carbon footprint relative to their cost. Buying a designer shirt, for example, may not seem like a particularly green thing to do. But given that most of the carbon footprint of a shirt is in growing and processing the cotton, and that a designer shirt is unlikely to use much more cotton than a cheaper one, then it follows that spending £75 on a luxury shirt will cause fewer emissions than buying two cheaper shirts with the same money. In other words, buying fewer but more expensive items is usually greener than buying a higher number of cheaper items.

Emissions per pound spent is a useful way of thinking about green purchases. Of course, the argument breaks down if you choose cheaper products in order to free up money to buy solar panels, loft insulation or other items that have a negative carbon footprint. In general, though, it's true that spending your money on fewer but more expensive products and services will help reduce your emissions.

Wooden products
Furniture, timber & other wooden things

Wood is a natural, renewable, recyclable, biodegradable and non-toxic material. It's more energy-efficient to produce than metal and most other materials and, in theory, if trees are planted in greater numbers than they're cut down, the timber industry can even help reduce global warming by soaking up CO_2. But some of the wood we buy comes with serious environmental and social costs. Every year, millions of acres of ecologically precious tropical rainforest is lost. Urbanization and the clearing of farmland for cattle feed and palm oil are probably the biggest drivers of this destruction, but the harvesting of valuable woods – such as mahogany and teak – is still a major contributor to the problem. Such woods are used in everything from salad bowls and garden furniture to musical instruments (see box on p.275).

As we saw in chapter one, deforestation is a key source of climate change. It's an enormous threat to biodiversity, too: more than half the world's species are thought to inhabit the three remaining large tropical forests, in the Amazon and Congo basins and the island nations of South-East Asia. The loss of forest also alters local climates, causes soil erosion and in many cases is associated with serious social problems. Many forest-based people have contracted diseases from loggers or been aggressively forced off their indigenous lands – or even killed when they've refused to leave.

The UK imports around four-fifths of its wood and when it comes to rainforest varieties such as mahogany, a large proportion of these imports have been illegally sourced – that is, cut down or exported in a way that breaks the laws of the country it came from. There are no definitive figures, but Friends of the Earth and other groups have estimated that more than half of the UK's tropical wood imports have been illegally sourced, with much of the rest being legal but not from sustainably managed forests. The timber industry disputes such claims, yet cases keep popping up which suggest that even wood specialists with ethical policies and big budgets are failing to keep their noses clean (as the Royal Family discovered when the Queen's Gallery was inadvertently refurbished with wood from endangered forests in Cameroon).

A recently cut rainforest tree being planked in the Peruvian Amazon

Of cues, caskets and clarinets

Furniture and timber are the products most widely associated with eco-unfriendly wood. But the same issues apply to tree-based products of all shapes, sizes and functions. For instance, as Andrew Wasley reported in *Red Pepper,* the UK imports nearly 300,000 budget snooker cues made "from the timber of the ramin tree – a rare species listed under the Convention on International Trade in Endangered Species (CITES) – chopped down and exported illegally from Indonesia's dwindling tropical forests".

Of the two hundred trees estimated to be used in musical instruments, meanwhile, more than seventy are included in the World Conservation Union Red List of globally threatened trees, according to Fauna & Flora International's SoundWood programme (www.globaltrees.org/soundwood.htm). Some of these so-called tone woods are trees that are critically endangered, including species of ebony and black wood used in several string and wind instruments, but not any longer on piano keyboards, Stevie Wonder songs notwithstanding.

Even as we leave the world we may not be treading lightly upon its forests, since some coffins are veneered or made out of tropical hardwood such as mahogany. (Many coffins also contain formaldehyde and plastic linings, which explains why crematoriums are responsible for a significant proportion of UK dioxin polution.)

The FSC logo may eventually become a common sight on all kinds of wooden products, but pressure from green-minded consumers will be required to speed this process along. Even just asking questions will help. As for alternatives, you're unlikely to come across cues or clarinets flaunting their green credentials. But a surprising number of environmentally friendly coffins are available from brands such as EcoCoffins.com and DaisyCoffins.com, in materials ranging from cardboard to water hyacinth.

Good wood

The best way to be sure that your wood has come from sustainably managed forests is to choose products bearing the logo of the Forest Stewardship Council (FSC). There are various sustainable wood labelling schemes out there. But many of them – such as the American Sustainable Forestry Initiative – are run by the timber industry and lack the credibility of the FSC.

An international, independent, non-profit-making organization, the FSC was founded in 1993 after extensive consultation between "timber users, traders and representatives of environmental and human-rights organizations". It only accredits wood when it can vouch for the entire supply chain – or chain of custody – from forest to sawmill to processor. The scheme has been criticized from time to time (in late 2002 the Rainforest Foundation accused the FSC of "knowingly misleading the public") but the vast majority of environmental groups fully support the organisation and its labelling scheme.

FSC-approved furniture, timber and wooden objects aren't difficult to come by. For a list of products and suppliers go to:

FSC Product List www.fsc-uk.org/product-search

Reclaimed wood

FSC-certified wood is a good choice, but environmentalists claim the very greenest option is to favour reclaimed (recycled) timber whenever possible. After all, for all its recyclable credentials, wood accounts for a significant proportion of our waste: Friends of the Earth has estimated that 3000 tonnes of perfectly good wood are thrown away or burnt each day just from buildings being demolished in the UK. And yet reclaimed wood is often better quality than new stuff, as it contains less water, is less likely to contract and was usually harvested before the advent of quick-growth forests, which are better at producing wood quickly than they are at delivering really fine timber.

The Reclaimed Building Supply website will help you find local suppliers of reclaimed timber and other products. Also check out Salvo, which has classifieds for everything from reclaimed floorboards to antique staircases, and the relevant page of the WasteBook site.

Bamboo

Bamboo is something of a wonder crop. A key building material in East Asia since time immemorial, it has also served as everything from a foodstuff to jewellery. Today it is hailed by some as a green alternative to hardwood. It's very strong yet doesn't contract as much as wood, and with modern processing it can even be used for flooring. It grows incredibly fast – some species can manage two feet in their first day – making it sustainable to grow in large quantities, unlike many hardwoods. Significantly, the way that bamboo grows means that when the plant is harvested it isn't killed: it simply grows back up from where you cut it, so the roots remain in place and the soil isn't damaged or washed away. Furthermore, bamboo doesn't require lots of pesticides, can grow nearly anywhere and is even thought to be capable of sucking pollutants out of the water cycle. All told, the material would be a contender for the world's most sustainable hardwood – were it not for the fact that it's actually a type of grass.

Reclaimed Building Supply www.reclaimbuildingsupply.com • 01883 346 432
Salvo www.salvo.co.uk
The WasteBook www.wastebook.org/furnrec.htm

New, uncertified wood

If buying FSC-certified or reclaimed wood isn't an option, at the very least try to avoid tropical or semitropical hardwood such as mahogany, teak, redwood, rosewood, ebony and iroko (an increasingly popular choice for high-end kitchen worktops). Such products won't necessarily have been logged from virgin rainforests, but it's not unlikely. Also try to avoid suppliers that can't tell you about the origins of their woods.

The chemical question
Toxic substances in household products

For some people, a key tenet of green living is the shunning of toxic or environmentally problematic synthetic chemicals in products as diverse as shampoo, detergents and paint. Indeed, environmentalists have been campaigning to reduce the use of toxic substances ever since the birth of the green movement. (A movement that emerged partly in response to the 1962 publication of Rachel Carson's *Silent Spring*, a book about the toxic chemicals used in farming.)

When it comes to climate change, household chemicals *aren't* a key issue. Indeed, it's probably the case that – as with plastic bags versus paper bags – many synthetic chemicals have a lower carbon footprint than the "natural" alternatives. For that reason, and because of space restrictions, this book has so far touched only fleetingly on chemicals. But it's worth quickly outlining some of the issues and listing some producers of non-toxic products.

The chemical world

As with climate change, the debate around chemicals and consumer products doesn't always focus on the most critical areas. One example of a chemical product that gets a lot of attention – possibly because of the ubiquity of green alternatives – is washing up liquid. That's a little ironic, because the detergents (or "surfactants") and other substances used in washing up liquid rarely feature on the lists of particularly harmful chemicals. At the levels used, they're not widely believed to be a serious health hazard to humans and, with water quality in UK rivers improving all the time, not everyone is convinced that they pose any serious risk to aquatic life. That's not to say that green washing up liquids are a waste of time; their plant-derived surfactants are definitely more planet-friendly than the petrochemical alternatives. But washing up liquid is not the key environmental area that some people believe it to be.

Rather than focus on detergents, it would make sense to take a broader view. Synthetic chemicals, including many that are far more problematic than the ones in our washing up liquids, are present in all kinds of products, from carpets (which often contain substances such as brominated flame retardants, which are believed to be dangerous to children) to PVC shower curtains and toys (which may contain phthalates, some of which are known to be potential hormone disrupters). Despite being in widespread use, some of these kinds of substances have been classified as "chemicals of high concern" by the EU.

Some chemicals, including the artificial musks widely used as fragrances in toiletries and laundry powders, are not just potentially toxic but also bioaccumulative, meaning they can build up in the body tissue of humans and other organisms, and be passed on through the food chain or by birth. "Babies are born with toxic chemicals already contaminating their bodies", according to Greenpeace, while wildlife groups have raised concerns that bioaccumulative chemicals are starting to turn up in the livers of arctic animals such as polar bears, presumably having been passed from products to people to water treatment systems, and then up the food chain via plankton, crustaceans, fishes and seals, increasing in concentration at each level. Chemicals can even pass back to humans via the consumption of fish.

So how serious are the environmental and health risks posed by these various kinds of chemicals? In most cases, the honest answer is that no one really knows for sure. In part this is because causal links between specific chemicals and specific effects are extremely difficult to prove. Whether you're talking about cancer in humans, or endocrine disruption in arctic birds, it's difficult to isolate and test the effect of any individual chemical. It's harder still to measure the possible "cocktail" effect of that same chemical consumed in combination with others.

Partly because of these kinds of challenges, there hasn't been a great deal of research into the impact of the thousands of chemicals used in everyday products. According to *Chemicals in Products*, a recent report from the Royal Commission on Environmental Pollution, "Society might

Corbis

Children dismantling European computers at a dump in Ghana, aiming to recover and sell materials. Green groups are lobbying electronics companies to phase out the most toxic chemicals, partly to avoid this kind of dangerous exposure.

reasonably expect that adequate assessments have been carried out on chemicals that are on the market, and that appropriate risk management strategies are in place for potentially harmful substances. This is not the case."

To better understand the risks to people and ecosystems posed by around thirty thousand chemicals already in use, the EU recently launched a major initiative called REACH ("Registration, Evaluation, Authorisation and Restriction of Chemicals"). REACH will test and monitor all the chemicals on sale in EU countries, including those arriving in the form of manufactured goods. The project has not been popular with some animal rights groups, since it is due to involve a large number of animal experiments, but anti-chemical campaigns see REACH as a huge victory – especially the "substitution" principle, which should require companies to phase out the use of chemicals of high concern whenever a safer alternative exists.

In the meantime, it's worth keeping the possible health risks in perspective. Even if certain chemicals are proved to be dangerous, it's likely that the risks will be many orders of magnitude lower than those associated with smoking – or even driving. Similarly, the environmental impacts of our use of chemical products is likely to be very small compared to the impact of our use of energy from fossil fuels. (That's true not only because fossil fuels cause global warming, but also because processing and burning gas, oil and coal causes large quantities of toxic chemicals to enter the atmosphere, from benzene to mercury.)

Of course, even if the risks to consumers are low, some synthetic chemicals could pose more serious threats to individuals working in the factories that produce them, or indeed to the people living in areas where toxic waste is dumped or stored. Having campaigned for years on chemical products in household produces, Greenpeace is currently focusing exclusively on the toxic substances in our gadgets, computers and other electronic items. This is an area of particular concern because discarded electronic equipment – so-called "e-waste" – is often dumped in developing countries, where some of it is even disassembled by hand.

Some ways to reduce your exposure to toxic chemicals

As we've seen, it's impossible to avoid all toxic or potentially toxic synthetic chemicals due to their sheer ubiquity. But individuals who are concerned about toxicity can certainly reduce their exposure by opting for household products that shun toxic substances in favour of gentler alternatives.

Cleaning products

Cleaner products based entirely or largely on plant-based chemicals of minimal toxicity are available from various brands such as Ecover, Bio-D and ACDO. Ecover in particular is widely available in supermarkets as well as health food shops, though for a full selection of products you might need to try a specialist green store such as:

Green Shop www.greenshop.co.uk
Natural Collection www.naturalcollection.com
Nigel's Eco Store www.nigelsecostore.com

If you feel particularly strongly about animal testing (or indeed you're a vegan and keen to shun all animal-based ingredients), you might want to seek out green products from a company approved by the very strict Humane Household Products Standard. These include:

Astonish (by the Oil Refining Company) www.astonishcleaners.com
Clear Spring (by Faith Products) www.faithinnature.co.uk

Some committed greens try to avoid conventional cleaning products wherever possible and to favour old-fashioned alternatives. Baking soda works well as a surface cleaner, as does lemon juice as a bleach substitute and white-wine vinegar solution as a descaler and glass cleaner. The more adventurous among the eco-minded community even swear by tomato ketchup as a pot scrubber.

Cosmetics & toiletries

A number of companies produce cosmetics and toiletries based on non-toxic and ecologically low-impact ingredients. Some also use organic ingredients and recyclable plastic bottles, and shun animal-based ingredients. Popular brands include:

Green People www.greenpeople.co.uk
Established in 1997, Green People sell more than one hundred organic (and vegan) toiletries and skin products, from sun creams to organic jojoba, hemp and rosehip oils. The company donates 10% of its net profits to environmental charities.

Honesty Cosmetics www.honestycosmetics.co.uk
Marketing their products "without unrealistic claims of emotional or physical benefit", Honesty Cosmetics sell a wide range of reasonably priced toiletries and cosmetics, including own-brand products made with natural oils sourced with environmental sustainability in mind.

Neal's Yard www.nealsyardremedies.com
Expensive they may be, but Neal's Yard products – some organic, all non-petrochemical and minimally packaged – are as good as toiletries get. Available in many specialist stores as well as online.

For yet more choices, try the following websites:

Culpeper www.culpeper.co.uk
Faith In Nature www.faithinnature.com
Weleda www.weleda.co.uk

Paints & DIY products

Though cleaning products and cosmetics get more attention, the chemicals in paint, brush cleaner, stripper, wood stains, glues, varnishes and other DIY products are often incomparably more toxic and ecologically problematic. In 1989, the International Agency for Research on Cancer went so far as to say that "occupational exposure as a painter is carcinogenic". Things have improved since then, as some of the worst chemicals have been removed from paints and other products, but many various highly toxic substances are still in use.

One issue is that many paints and solvents – in both their manufacture and application – release volatile organic compounds (VOCs) into the air, leading to the creation of polluting ground-level ozone, among other things. The EU has tightened up the laws about VOCs, and paint manufacturers now specify the VOC level on their tins. This development has been welcomed by chemicals campaigners, but it doesn't deal with the heavy metals, solvents and other controversial chemicals used in many paints and other DIY products. From dyes to plasticizers, many ingredients are fat-soluble and therefore prone to accumulating in the bodies of people and animals.

Paints and varnishes can continue to emit fumes once they're applied and this is thought to be a contributor to sick building syndrome (SBS) – headaches and other health effects which seem to be brought on by a specific building. Though this all sounds a bit New Agey, SBS is an all-too-real malaise, formally recognized by the World Health Organization since 1982.

As for the effect on the wider environment, the amount of VOCs, bioac-cumulative toxins and other chemicals released into the atmosphere at the time of application is relatively small with most DIY products. That is, as long as you don't pour any down the sink (instead, give unwanted half-full paint tins to charity; see p.123). But, according to environmental groups, the manufacture of a litre of paint can create ten or more litres of toxic waste – which, considering that the UK gets through hundreds of million of litres of paint each year, adds up to a staggering total.

As a general principle, products that are water soluble are likely to be less toxic and certainly less bioaccumulative than those based on solvents. So it makes sense to favour water-based gloss, for example, which is available in all good DIY shops. For maximum eco-friendliness, though, opt for products from a specialist green brand. The following companies produce a wide selection of paints as well as some varnishes, strippers, waxes, stains and other related products.

Auro Organic www.auroorganic.co.uk
ECOS www.ecospaints.com
Ecotec Paints www.natural-building.co.uk
Livos www.livos.demon.co.uk
Nutshell Natural Paints www.nutshellpaints.com
OSMO www.osmouk.com

You can order colour charts and buy online from most of these suppliers, but for advice you might be better contacting a specialist store, such as:

Construction Resources (London) www.ecoconstruct.com • 020 7450 2211
Eco Merchant (Kent) www.ecomerchant.co.uk • 01795 530 130
Green Building Store (West Yorks) www.greenbuildingstore.co.uk • 01484 854 898
The Green Shop (Gloucestershire) www.greenshop.co.uk • 01452 770 629
NBT (Buckinghamshire) www.natural-building.co.uk • 01491 638 911

For some DIY products, there's no particularly green equivalent. For example, the Association for Environment-Conscious Building recommends avoiding wood preservative altogether. If it must be used, they claim, boron (available from Auro and Livos) is the best option.

Other items

In addition to those listed above, products specially designed to reduce exposure to toxic chemicals include organic food (see p.183) and organic clothes and fabrics (discussed below). There are also various companies selling toys made from natural materials such as wood, reflecting the fact that many chemicals are thought to pose greater risks to children than they do to adults.

Holz Toys www.holz-toys.co.uk
Myriad Toys www.myriadonline.co.uk

For yet more low-toxicity items, browse the follow green supersites:

Green Shop www.greenshop.co.uk
Natural Collection www.naturalcollection.com
Nigel's Eco Store www.nigelsecostore.com

Clothes & fabrics
Clothes, shoes and soft furnishings

As we saw in chapter two, our clothing accounts for almost a tenth of our carbon-dioxide emissions. Perhaps surprisingly, the majority of those emissions are caused not by producing the clothes themselves, but by generating the electricity to power the machines we use to wash and dry them. Probably the single most effective way to reduce the carbon emissions of your wardrobe, therefore, is to use short, low-temperature washing cycles and to favour a clothes line or rack over an electric dryer – as discussed in chapter six.

As for buying the clothes in the first place, perhaps the most important thing is to choose items that are likely to last well and to look after them properly. But it's also worth considering the materials used to produce the clothes and shoes that you purchase. Here's a quick look at some of the most popular materials.

Natural fibres: cotton, wool and hemp

Cultivated by humans for more than 5000 years, and the basis of products ranging from T-shirts to US dollar bills, cotton is something of a natural wonder. But much has changed since the ancient Greek historian Herodotus first documented the existence of "tree wool … exceeding in beauty and goodness". Today, cotton lies at the heart of global debates on agrochemicals, GM versus organic farming, and ethical versus regular trade. The agrochemical question relates to the fact that cotton requires more pesticides than almost any other crop. Despite covering only a few percent of the world's cultivable soils, cotton farms account for roughly 10% of all herbicide use and an astonishing 20–25% of insecticide use.

People argue about whether or not these chemicals – which include carcinogenic and otherwise nasty substances such as carbamates and organochlorines – are bad for the eventual wearer of the cotton. But the effects on farmers and the environment are more pressing. There are no definitive global figures but anti-pesticide campaigners claim that thousands of cotton-farm workers are killed by agrochemicals each year. Research by Pesticide Action Network, for example, showed that in a single region within the tiny West African state of Benin "at least 61 men, women and children" were killed by cotton pesticides between 1999 and 2001. And, of course, for every one person who dies, hundreds of others have to put up with the ill effects of polluted air and water.

Some commentators also worry that the chemicals are entering the food chain, as the vast quantities of cottonseed harvested are fed to animals and made into cottonseed oil for human consumption.

There are two proposed solutions to the cotton chemical problem. Biotechnologists advocate using GM cotton crops to cut back on pesticides, while greens advise consumers to opt out of pesticides *and* GM by choosing organic cotton or favouring alternatives such as hemp (see box). Just as with food, organic fibres can be grown using chemical-free alternatives to pesticides. In Uganda, for example, black ants are

Hemp

While some people advocate organic or GM cotton as the greenest of fabrics, others sing the praises of hemp – the non-psychoactive industrial counterpart of marijuana. The fact that most of hemp's rather evangelical advocates also seem to be keen smokers of their favourite plant doesn't do the campaign many favours. But with increasing numbers of governments and farmers examining its benefits, it seems they may have a point. *Cannabis sativa* has a long and distinguished history, taking in at least 5000 years of use as a source of food, cloth, nets and countless other items. Many products have started out life as hemp, from proper paper – first created in Tibet at roughly the time of Christ – to Levi's jeans. In the 1940s Henry Ford famously produced a partly hemp-resin car powered by hemp-based fuel.

Until the twentieth century, the plant was also the main source for ropes and sails (the word canvas is derived from the Latin *cannabis*), which made it so central to naval success that both Britain and the US once had laws requiring all farmers to dedicate a proportion of their land to it. However, a couple of centuries after George Washington advised Americans to "make the most of hempseed and sow it everywhere", the wonderplant was on the decline, gradually being replaced by cotton for clothes, wood for paper and nylon for rope. Once the first round of the war on drugs got under way, things got even worse, with hemp officially or effectively banned in many countries.

Hemp fans claim that now is the time for a revival. The environment would benefit through reduced chemical use, improved soil structures (due to the plant's deep roots) and reduced felling of trees for paper and board, since hemp fibre is far quicker to produce at a sustainable rate than wood fibre and requires less energy input and chemical treatment. Farmers would gain because the plant is unusually easy, quick, inexpensive and safe to grow. To cap all this off, the plant's seeds are unusually nutritious, with a blend of amino acids close to perfect for humans. In remains to be seen whether a real hemp revival will ever materialise, but a wide range of products are already available from the eco clothes brands listed overleaf and specialists such as braintreehemp.co.uk.

collected and set on the cotton fields to eat the pests. Organic cotton is also more eco-friendly in terms of processing. In the manufacture of conventional cotton cloth thousands of synthetic substances are used: chlorine bleaches, heavy-metal dyes and treatments such as formaldehyde to lessen creasability and minimize shrinking. Many of these are highly toxic, and can cause environmental and health problems if not handled and disposed of correctly.

The market for organic cotton is rocketing, thanks to orders from specialists and from a few big companies – including Nike and Levi – who buy a small proportion of their cotton from organic sources. The US used to be the biggest producer, but increasingly organic production is spreading to developing countries, which can undercut the West when it comes to low-chemical, labour-intensive production methods.

Organic cotton, then, may have some social as well as environmental benefits. But it doesn't directly address broader social issues such as the fact that millions of developing-world cotton growers – like coffee farmers – have to contend with fluctuating commodity markets and unscrupulous middlemen. There's also the question of child labour. According to a study by the India Committee of the Netherlands, 90% of all labour in the Indian cottonseed market is carried out by nearly half a million children, mostly girls aged between six and fourteen. It's these kinds of social issues that led to the emergence, in 2005, of Fairtrade-certified cotton. This works in a very similar way to other Fairtrade crops (see p.205), guaranteeing the farmer a minimum price in addition to a social premium to invest in community development projects. The system also requires certain minimum labour standards in the factories that produce clothes from Fairtrade cotton, though the primary focus is firmly on the growers.

After cotton, the most popular natural fibre for making clothes is wool. Wool can be produced with relatively few agrochemicals, though from a climate change perspective it probably has a relatively high carbon footprint due to the methane and nitrous-oxide emissions of the sheep and their manure (see p.213). It's hard to say with any precision how wool compares to other materials, since little if any research on this question has been published. If lamb and mutton are anything to go by, though, it seems likely that organic wool will be more carbon-friendly than wool from a conventional farm.

Eco clothes companies

There are now a large number of companies selling clothes, bedding and other items made from organic fibres or other materials selected for their eco-friendliness. These include:

Gossypium www.gossypium.co.uk • 0800 085 6549 • Abinger Place, Lewes
Elegant, simple T-shirts, tops, night-wear and knickers – plus yoga and kids' clothes – in Fairtrade organic cotton from India.

Greenfibres www.greenfibres.com • 01803 868 001 • 99 High St, Totnes
Organic clothes for men, women and children – underwear, casual wear and even suits – as well as towels and sheets and other home furnishings.

Hug www.hug.co.uk • 0845 130 1525

Nicely cut T-shirts, tops and jeans for men, women and children. The style is Gap-esque but the cotton is Fairtrade and produced with eco-friendly dyes. Items arrive in a letter-box-friendly parcel; if you don't like it, simply send it back freepost.

Howies www.howies.co.uk • 01239 615 988

Eco clothes in a surfer/skater style, using organic natural fibres as well as recycled polyester.

Patagonia www.patagonia.com/europe • 01629 583 800

High-quality fleeces, jackets and other outdoor-pursuits gear made from organic cotton by a company with good environmental credentials.

People Tree www.ptree.co.uk • 0845 450 4595

Selling online and in certain Top Shop stores, People Tree offers one of the biggest ranges of fair trade clothes, mostly in organic cotton.

Natural Collection www.naturalcollection.com • 0870 331 3333

A full range of men's and women's clothes (casual wear, sportswear, knitwear, under-wear and more) in organic cotton and hemp, as well as bedding, towels and other homeware.

Ralper www.ralper.co.uk

T-shirts with and without screenprints, branded to appeal to a young market. Garments are Fairtrade-certified and mainly in organic cotton.

Leather and alternatives

Leather may be a "natural" product, but it is associated with at least two environmental problems. First, although leather is often described as a byproduct of meat farming, the hide of an animal adds substantially to its slaughterhouse value. In other words, the demand for leather subsidizes meat production and the leather trade is partly responsible for the massive environmental impacts of livestock discussed in chapter fourteen. A three-year project by Greenpeace in Brazil recently traced leather from cattle raised on freshly deforested ranches in the Amazon and found it in use by Western brands ranging from Adidas and Nike through to BMW and IKEA.

The second environmental concern about leather is the fact that modern-day tanneries make use of various chemical processes, some of them involving highly toxic substances such as salts of chromium. This processing not only makes leather surprisingly energy-intensive to produces but also produces massive quantities of waste: treating a ton of raw hide can result in 75,000 litres of waste water and 100kg of dried sludge.

As environmental regulation has tightened up in the West, developing-world tanneries have seen their businesses expanding. This provides much-needed jobs and income, but the human and environmental costs can be high. A study by the Tokyo Institute of Technology found that 62% of the 2.4 million people in Kasur, Pakistan, suffer from ailments – including some as serious as cancer, tuberculosis and blindness – caused by industrial waste, with tanneries the major contributor. Similarly, in the Bangladeshi capital Dhaka, according to a report cited in the *World Health Organization Bulletin*, half a million residents are at risk of serious illness due to tannery pollution, with tannery workers there 50% more likely than their peers to die before the age of fifty.

You may occasionally come across leathers tanned and dyed with natural substances but this is still something of a niche. Moreover, natural leather treatments may include palm oil, which is linked to rainforest destruction. For all these reasons, some people advocate "vegetarian" leather substitutes. Being made of polyurethane and other plastics, these aren't exactly the greenest products in the world themselves, but they're arguably lower-impact than leather. Suppliers include:

Beyond Skin www.beyondskin.co.uk • 0845 373 3648
High-end non-leather shoes for female fashionistas. Customers include the likes of Natalie Portman.

Vegan Store www.veganstore.co.uk • 01273 302 979
A range of animal-free shoes and boots as well as belts, wallets and even watch-straps.

Vegetarian Shoes www.vegetarian-shoes.co.uk • 01273 691 913
An impressively wide range of leather-free footwear, including boots, smart shoes, trainers and Camper-style day-to-day wear.

Money matters

Green banking and finance

Most of us trust our money to institutions that attempt to increase its worth – for our gain and their own – by investing it in shares, bonds and property, loaning it to companies, countries or individuals, or gambling it on currency or commodity markets. This goes not only for the cash in our bank accounts, but also for our pension contributions and investments as well as the money we give to insurance companies.

Whether that's a problem depends on whether you care who your money is invested in or loaned to. Unless the financial institution in question has an investment policy stating otherwise, your cash could end up supporting environmentally or ethically problematic projects – anything from the building of coal power stations through to arms sales to oppressive governments.

These issues are particularly pertinent because financial institutions are so powerful. As the recent economic downturn amply demonstrated, irresponsible behaviour in the financial sector can have tremendous impacts on all aspects of society. Even in economically happier times, pension funds and insurance companies control more than half of the UK stock market, so their investment decisions are hugely significant in terms of the direction of financial flows.

In response to increasing public awareness of these kinds of issues, a growing number of banks, pension funds and other bodies are offering green and ethical financial services. This chapter provides some context and then gives specific information about banks, pensions, investments, insurance and mortgages.

Green money: the basics
How socially responsible investment works

The basis of most green and ethical financial services is socially responsible investment, or SRI, the practice of considering environmental and social factors when deciding which shares to buy and who to loan money to. This concept is nothing new – it can be traced back at least a hundred years to the time when Quakers and Methodists started boycotting certain sections of the stock market. But in the last decade SRI has become a booming sector. There are now a number of banks, many pensions and more than fifty UK investment funds driven by SRI policies, plus numerous financial advisers specializing in the field.

The green spectrum

Ethical investment policies vary widely, from the highly strict – informally dubbed "dark green" – to the more flexible, or "light green". Dark-green investment policies are traditionally based on negative screening. This involves drawing up a list of unacceptable practices deemed to be harmful to people, the environment or animals and excluding any company found to be involved in these practices. An investment organization might decide upon its own criteria, or it could get some pointers from an external body, but either way the approach involves excluding entire industries – nuclear power, arms manufacture and tobacco being a few common examples – as well as any individual company associated with unethical behaviour.

Another dark-green approach is so-called cause-based investment (also known as alternative investment, mission-based investment and socially directed investment). Practised by the more specialist ethical organizations such as Triodos Bank (see p.297), this involves investing directly in projects and companies deemed to have social and environmental worth, usually completely avoiding loans to big companies and investments on the stock market. Typical beneficiaries are charities, organic farms and community housing projects – the kinds of organizations that might struggle to find the credit they need at affordable rates elsewhere.

Light-green investors take a different approach. They may rule out a few industries, such as arms manufacture, but generally they prefer positive screening: any company can qualify for investment as long as it fulfils a certain number of positive criteria, such as full recycling of waste or the publishing of in-depth, publicly available reports on the company's environmental impact. A variation on this approach is best-of-sector screening, which involves investing only in the most ethical company in each sector. The logic is simple: if you accept that, say, nuclear power companies are inevitably going to exist, giving them all an incentive to be the best in the field may make a bigger difference than boycotting the whole sector.

In addition to screening, light-green lenders and investors might attempt to use their insider influence to push for better standards. For example, a pension fund which owns shares in a supermarket chain might threaten to cause a fuss at the annual general meeting unless it agrees to improve its environmental performance.

Common screening criteria

Following are the kinds of pluses and minuses that an ethical finance organization might be looking out for:

Negative

▶ Industries The blacklist often includes alcohol, animal experiments, arms, fur, gambling, genetic engineering, intensive farming, nuclear power, oil, pornography and tobacco.

▶ Environment Association with specific problems such as chemical pollution, deforestation or carbon-intensive products; or a simple lack of environmental policies.

▶ Human rights Lack of a code of conduct on workers' rights; association with human rights abuses of any kind; or links to oppressive governments.

▶ Management Excessive directors' pay; lack of financial transparency; political donations; conflicts of interest; use of tax havens.

Positive

▶ Charity & community Charitable donations; participation in and support for local events; sensitivity to the business's effect on local people.

▶ Environment Clear environmental policies; environmental auditing and reporting; minimizing pollution and energy use.

▶ Staff Clear codes of conduct on pay and labour conditions, equal opportunities and staff "development".

▶ Management Disclosure of payments to foreign governments; compliance with "corporate governance" protocols.

Dark green vs light green

The dark green vs light green argument largely comes down to the old question of idealism vs pragmatism. Advocates of the dark-green approach think that it's morally unacceptable to profit from any companies or industries whose activities may cause harm to others. Some even claim that the lighter-green banks and investment funds actually do harm – by giving the thumbs up to dodgy firms, and in the process helping them create a veneer of social responsibility.

On the other hand, the pragmatists argue that working *with* companies – through engagement and positive screening – is more likely to make a difference than avoiding their whole sector. It's better to have progressive bankers and investors involved in every industry, they claim, than to leave the most unscrupulous financial backers and most dodgy businesses to get on with wrecking the planet. After all, dark-green ethical banks, pensions and investment funds have a tiny market share (less than 2% in the UK), and there are plenty of other lenders and investors queuing up to finance or buy into even the murkiest companies.

Furthermore, light-green advocates point out that there's not much logic to financially boycotting sectors which we continue to support as consumers and voters. Does it make sense, for example, to object to our money being invested in oil companies while we continue to drive cars? Or to shun arms firms unless we're committed to the total abolition of all armed forces?

These are all fair points. But does the pragmatic option actually work? Can investors and bankers really make a difference by "engaging" with companies? Exponents of this approach claim there have been numerous successes. Friends of the Earth point to McDonald's phasing out environmentally problematic polystyrene packaging, for example, and Ford pulling out of the Global Climate Coalition (a now-defunct pressure group which argued against the Kyoto treaty and other measures to combat climate change). Others point to the role that shareholders played in the fall of apartheid in South Africa. And British ethical fund managers claim to have made big impacts behind the scenes on a whole range of issues – including the introduction of carbon reduction targets and codes of conduct governing labour abuses in overseas garment factories.

Despite all these examples, there are few cases where it can be said definitively that shareholders' ethical concerns have resulted in a company changing significantly for the better. That's perhaps not too surprising, because the financial "engagers" are only likely to change a company's practices if they can show not just a moral case for improving behaviour, but also a financial one. Sometimes this may be possible, but very often it's simply not true that better corporate behaviour means more profits. As one city analyst told *The Observer* back in 2002, "On current share trends it pays to be socially irresponsible all the way."

To really *force* a company to change requires the progressive investors – or shareholder activists, as the more extreme ones are known – to table a resolution at the company's AGM (annual general meeting). These are occasionally successful on some issues, such as the 2003 GlaxoSmithKline shareholders' protest against the outrageous pay awarded to the company's CEO. But specifically green or ethical resolutions are rare in the UK, and when they do happen, they don't generally achieve landslide support. The Greenpeace-led resolution against BP Amoco's Northstar project in Alaska, for example, was considered a major success when it achieved around 13.5% of the vote.

But that doesn't mean the engagement approach is worthless. Such resolutions can force issues into the AGM as well as the media. They can also play their part in wider protest. For instance, British construction company Balfour Beatty withdrew from the Ilisu Dam project in Turkey which would allegedly have displaced thousands of people and have had potentially disastrous environmental consequences – after a wide-ranging campaign against the company. Shareholders were not solely responsible, but they played their part, delivering, in the words of Simon McRae from Friends of the Earth, "a big slap in the face" at the AGM.

Ultimately, dark-green and light-green strategies are both valid, and in practice many banks and investment funds favour a mixed approach: excluding certain industries and giving preference to companies with positive practices; but also taking the engagement path on some issues. Similarly,

Corbis

Environmental protesters lobbying Exxon shareholders outside the company's annual general meeting

rather than boycotting a whole industry, an organization may be more selective – the Co-operative Bank policy, for instance, doesn't categorically rule out all companies involved in manufacturing armaments for defence, but it won't invest in any firm which produces torture devices, or any which exports arms of any kind to countries deemed to have oppressive regimes.

Green money and interest

In the case of banks, having a green or ethical account certainly doesn't need to mean you being poorer. Most of the population get virtually no interest on their high-street current account, and pay over the odds for services and overdrafts. By contrast, Smile, the Internet arm of the environmentally and ethically progressive Co-operative Bank, offers low fees and excellent interest rates. (The Smile current account rate has shrunk to almost zero with the recent slashing of Bank of England base rates, but should rise again in due course.)

Things are a bit more contentious when it comes to investments and pensions, depending on how long a view you're prepared to take. Over the really long term, the evidence seems to suggest that "screened" pension and investment funds are actually just as good a bet as non-ethical ones. Probably the most in-depth assessment of the issue is *Does Ethical Investment Pay?*, produced in 1999 by the Ethical Investment Research Service (EIRIS). The study examined the performance of fifteen leading ethical funds over a long period, and found that ethical investment generally involves marginally less risk, but offers marginally lower returns than conventional investment. The balance of risks and returns, it concludes, is "not materially different". In other words, you're less likely to see your savings rocket, but also less likely to see them plummet.

Since then, some studies – such as an Australian survey carried out by AMP Henderson – have even suggested that ethical investment funds are *more* profitable than average. But things have looked rather different since the world's economy ground to a halt in 2008. Because green and ethical funds don't tend to include virtually recession-proof stocks such as tobacco, and because they often invest in small companies that are prone to credit squeezes, they have not been particularly good at weathering the economic storm. A study published at the time of writing by Moneyfacts.co.uk showed that ethical funds had taken a 30% hit since mid-2008, compared to a 25% hit for conventional funds.

Who gets the money?

It is sometimes said that "ethical" banks, pension managers and investment funds do invest in dodgy companies – just slightly less dodgy companies than the alternatives. It's certainly true that most SRI funds, while they shun arms, nuclear power, tobacco and the like, nonetheless invest in a list of companies that you might not exactly think of as moral trailblazers. As the box below shows, the most popular shares in ethical portfolios include major banks, supermarkets and retailers that are themselves often criticized on environmental and other ethical grounds.

This may not bother you, but if it does, bear in mind that not all ethical policies are alike. The Triodos Bank, for example, only finances schemes and companies "which add social, environmental and cultural value" to the world; and there are investment policies out there ranging from vegan (avoiding all forms of animal products and testing) to Islamic (ruling out alcohol and money-lending firms), so there's no reason why you shouldn't find one that suits you. Ask a specialist financial adviser for more information. To find one, see p.302.

The ethical share chart

The following list, drawn up by EIRIS in 2002, shows the companies whose shares crop up in the biggest number of ethical investment funds – starting with the most popular. There isn't an up-to-date version, but while individual companies will have come and gone, the type of firms represented is likely to have changed relatively little.

1 Vodafone Group	17 Cable and Wireless
2 FirstGroup	18 mmO2
3 National Express Group	19 Northern Rock
4 Prudential	20 CGNU
5 HBOS	21 Electrocomponents
6 Abbey National	22 The Sage Group
7 BG Group	23 Pearson
8 Centrica	24 Tesco
9 ARM Holdings	25 Xansa
10 Halma	26 Reed Elsevier
11 The Royal Bank of Scotland Group	27 Compass Group
12 Berkeley Group	28 Johnson Matthey
13 BT Group	29 Debenhams
14 Reuters Group	30 First Technology
15 RPS Group	31 Marks & Spencer Group
16 Nestor Healthcare Group	32 The Go-Ahead Group

Banks & building societies

Alternatives to the major banks

The major banks have been accused of all types of bad practice over recent years. Environmentalists charge them with funding deforestation, ecologically destructive infrastructure projects and polluting companies; anti-poverty groups accuse them of wrecking developing economies by gambling on currency values; even the Competition Commission has accused them of making too much profit. And that was all *before* their role in the 2008 economic crisis transformed them into by far the least popular brands on the high street.

Ironically, perhaps, the crash happened just as the major banks were starting to claim progress on environmental and other ethical issues. In the run-up to 2008, most of the leading banks had made some headway on reducing their operational carbon footprints by reducing the energy use in their offices, offering paperless accounts, and so on. And though none of them had yet offered anything resembling a robust green and ethical investment policy, they had at least signed up for the Equator Principles, a set of basic guidelines on socially and environmentally sound financing.

That's not to say the banks were or are on a truly green path. Campaign groups such as Bank Track argue that the Equator Principles have done little to stop banks making loans and offering accounts to companies involved in environmentally and ethically problematic activities. And according to industry insiders any hope of more meaningful green and ethical agendas has fallen away since 2008, as the banks have focused increasingly on simply staying afloat.

If you'd like to switch your current account or move your savings to a more ethically progressive organization, explore the following companies.

Green/ethical specialists offering current accounts

The Co-operative Bank www.co-operativebank.co.uk • 08457 212 212

Set up in 1872 as the Loan and Deposit Department of the Co-operative Wholesale Society, the Co-operative Bank now has more than three million customer accounts and is the only UK clearing bank to publish an ethical investment policy – something it first did in 1992. The policy, which is determined through regular consultation with the bank's customers, addresses concerns about human rights, arms, corporate responsibility, genetic modification, social enterprise, ecological impact and animal welfare. A mix of negative and positive screening, it's a medium shade of green:

companies that use animal tests are okay, for example, but only in the field of medicine; arms firms that export to oppressive regimes are no-go, but defence firms in general aren't explicitly ruled out. The bank also attempts to support charities and positive business ventures such as fair trade.

Practicalities: More than 300 branches and Handybanks; pay in and withdraw at post offices; 30,000 Link cash machines; 24-hour phone and Internet banking. Accounts offered: current, numerous savings, ISAs, business, student.

Smile www.smile.co.uk • 0870 (THE BANK) 843 2265

The UK's first Internet-only bank, Smile has been a huge success since it was launched as the online wing of the Co-operative Bank in 1999. It shares the same ethical policy as its parent but as it has no branches it can offer better interest rates. It regularly scores highest in the country in banking customer-satisfaction surveys.

Practicalities: No branches; excellent Internet banking with 24-hour backup phone line; postage-free address for sending cheques; pay in/withdraw at post offices; 30,000 Link cash machines. Accounts offered: current, instant savings, ISAs, student.

The Ecology Building Society www.ecology.co.uk • 0845 674 5566

Established in 1981, the Ecology is a small mutual building society whose savings accounts are only used to fund mortgages considered to be green (renovation projects, energy-efficient construction and the like). It's perhaps the world's only financial organization that will enquire if you are a member of a green group before offering you an account.

Practicalities: No branches; pay in/withdraw by cheque or transfer. Accounts offered: instant and sixty-day-notice savings accounts; Cash Mini ISA.

Green/ethical specialists offering savings accounts only

The Triodos Bank www.triodos.co.uk • 0500 008 720

Established in Holland in 1980, with a British office opening in 1995, the Triodos Bank is a specialist ethical savings bank with a "cause-based" approach. It only invests in "organizations which create real social, environmental and cultural value" – such as charities, environmental initiatives, social enterprises and community projects – but it still offers good rates of interest. Triodos doesn't offer current accounts, except for green businesses and charities, but it does offer a wide range of savings accounts focusing on investment in a more specific area: Charity, Just Housing, Earth, Organic, etc. Notice periods are one, thirty or ninety days. A model of transparency, the bank publishes a full list of all the companies and organizations it invests in.

Practicalities: No branches; pay in/withdraw by cheque or transfer. Accounts offered: ISAs, young saver, regular saver, social investor cheque, and others.

Charity Bank www.charitybank.org • 01732 520 029

Describing itself as "the world's first not-for-profit bank", Charity Bank was established in 2002 by the long-standing Charities Aid Foundation, an organization that provides credit to charities, which frequently find credit difficult to come by. The bank itself is a registered charity and its notice-period savings accounts are aimed at people more interested in knowing that their money is doing something useful than in earning huge returns. However, over and above the interest on offer, customers can make substantial savings on their tax bill by investing money for a minimum of five years. For high-rate taxpayers, the bank claims, this can work out at a return equivalent to more than 8%.

Practicalities: No branches; pay in/withdraw by post. Accounts offered: various notice-period savings.

Mutual building societies

Though they don't really fit into the green specialists category, mutual building societies (ie those which haven't been turned into banks) are often described as a more ethical place to put your cash than the high-street banks. After all, they don't invest in anything other than people's mortgages and, since they don't take business accounts, they don't have any environmentally dubious multinationals as customers. Most mutuals also make significant policy decisions democratically, with each saver entitled to one vote, regardless of the amount in their account.

In addition, mutuals generally offer very good rates of interest – perhaps because they don't have shareholder dividends to pay, and perhaps because their focus on service rather than profits allows them to be much more efficient (running costs rise about 35% when mutuals turn into banks, according to the Building Society Association).

Building societies that are no longer mutuals include Abbey National, Alliance and Leicester, Bradford & Bingley and Halifax. For a full list of the remaining mutuals, and for more information, visit:

Building Society Association www.bsa.org.uk

Credit unions

Credit unions offer a cooperative-style alternative to a regular savings account or investment vehicles. The idea is that a group of people who share a common bond – such as a residential area, employer, occupation or church – join together to form a union. They invest their savings into a pool, from which loans can be made to members.

Credit unions claim to offer ethical and financial benefits over conventional savings options. On the ethical level, the unions can help tackle financial exclusion. Banks often refuse finance to the poorest members of society due to their having a credit history tarnished long ago, but credit unions can make a case-by-case decision on who can borrow based on their current ability to repay and savings record with the union. Furthermore, since only members can borrow, there are no loans to environmentally problematic businesses. On the financial level, savers can expect dividends of up to around 8% on their investment (largely because there are so few overheads to pay), and borrowers receive better terms on their loans than a bank would offer. For example, there are no arrangement or early redemption fees.

Credit unions are owned by their members and are run democratically within a clear legal framework, including the obligatory training of people elected as officers. They've been around since the nineteenth century and have grown to include a hundred million members in 35,000 unions in more than eighty countries (though the majority are in the US and Canada). For more information, or to locate a credit union, visit:

British Association of Credit Unions Limited www.abcul.coop

Investments
Pensions, funds and financial advisers

Pensions

It's difficult to overestimate the financial weight of pension funds. In the UK they're worth hundreds of billions of pounds and own around a third of the stock market. As such, they have the potential to put enormous pressure on companies to improve their environmental and ethical standards. This potential is a very long way from being fully realized, but the last decade has seen a steady rise in the number of pension funds taking ethical factors into consideration when investing. One reason is a law passed in July 2000, obliging all trustees of stakeholder and occupational pensions to declare if and how "social, environmental or ethical considerations are taken into account in the selection, retention and realization of investments". The trustees are not obliged to invest ethically, but they must at least state their position and make it available to all members in a document called a Statement of Investment Principles (SIP).

This is all good news, but research shows that there is still a big gap between what pension-fund members want and what is actually happening. According to studies carried out by EIRIS and SustainAbility, the vast majority of people think their pension fund *should* operate an ethical policy, give members input into it and publish a list of companies invested in. But, as with banking, only a handful of the major funds have taken SRI principles to heart. Campaign Against the Arms Trade highlighted this fact in 2007, when they published lists showing the number of shares in arms companies owned by the major pension schemes. NHS trusts, trade unions and nearly every local authority were all revealed to have defence firms in their investment portfolios – so it seems fairly safe to assume that they have ecologically destructive companies in there too.

Greener pensions

If you already pay into an occupational or stakeholder pension, you can request the SIP from your fund managers, asking for clarification if the wording is unclear or noncommittal. It may also be worth searching on the Internet for impartial information about your scheme (for example, to see whether it's listed in the arms-trade study mentioned above). Once you have the information you need, if you're not satisfied with the stance of your scheme, send your views to the managers – and encourage like-minded colleagues to do the same. You may want to lobby for the intro-duction of a specific ethical plan, or request that social or environmental

Corbis

Many banks and pension funds have come under criticism for investing in companies linked to rainforest destruction. Deforestation exacerbates climate change as well as species loss. The Sumatran tiger has been pushed to the brink of extinction by a combination of forest destruction and hunting.

considerations are taken into account on the standard plan. If you want help working out what to ask and say, visit:

Fair Pensions www.fairpensions.org.uk

If all else fails, you could ditch your employer's scheme and switch to a separate stakeholder or personal pension plan that reflects your environmental or ethical criteria. However, this may significantly reduce the amount you are able to save (you'll probably lose any employer contributions) so you'd be well advised to think very carefully before making any decision, and to seek advice from a financial adviser before proceeding. See p.302 to find an ethically enlightened one.

If you have a personal pension (ie a non-employer pension that's not a stakeholder), your fund isn't obliged to inform you of their angle on responsible investment. However, it is highly likely that they will be happy to reveal their position and also offer you the option of transferring to a specific green or ethical plan.

If you're feeling particularly vigilant, you may also want to contact your council. As Martin Hogbin from Campaign Against the Arms Trade points out, council taxpayers also have the right to "question the investment policies of their local authority, as their taxes are used to top up existing pension funds".

Retail investment funds

If you have money to invest separately from your pension, you could choose to put it in one of the various savings accounts already discussed, or to join a local credit union. Alternatively, you could invest it in an SRIF – a socially responsible investment fund – of which there are now more than fifty in the UK. These funds run the gamut of ethical investment strategies described at the start of this chapter and offer products ranging from unit trusts to investment trusts and ISAs. As of 2007, they have an estimated total value of more than £8 billion – up from just a few hundred million pounds at the beginning of the 1990s.

With so many options out there, and new players entering the market all the time, a list of funds here would be of little use. Instead we've provided links to resources where you can find information. But the best idea is to speak to an ethically enlightened financial adviser, as described overleaf.

EIRIS Green & Ethical Funds Directory www.eiris.org
Published by the Ethical Investment Research and Information Service, *The Green &*

Ethical Funds Directory contains summaries of each fund's policies and gives their top ten holdings. It's available to download for free.

Ethical Investors Group www.ethicalinvestors.com • 01242 539 848

The Ethical Funds Directory on the website of this group of independent financial advisers provides basic information about the policies of many UK ethical funds. The site also covers general issues about ethical finance and provides some information on mortgages, pensions and other areas.

The Green Investment Guide www.thegreeninvestmentguide.com

If you're keen to invest in clean technologies that could help combat climate change – anything from wind and solar to energy storage and tidal power – then this site offers all kinds of useful background information and company snapshots.

SocialFunds.com www.socialfunds.com

This is a US site, so not all the information is relevant to UK readers. But it constitutes a massive resource and includes news on global SRI developments.

Trustnet www.trustnet.com

This massive, free-to-access website provides current and past performance information about all kinds of investment funds. Click Ethical/Sustainable in the Fund Focuses dropdown to see the green funds.

UK Social Investment Forum www.uksif.org • 020 7405 0040

Established in 1991, the UK Social Investment Forum is a "membership network" for promoting socially responsible investment. The Member Directory provides links to nearly all the main companies and organizations involved in ethical investment, and the homepage covers recent news and developments.

Independent financial advisers

An increasing number of financial advisers, or IFAs, are developing expertise in the green and ethical sector, so finding advice shouldn't be difficult. The best way to locate an ethical specialist near you is via the Ethical Investment Association, the UK Social Investment Forum or the Ethical Investment Research Service.

Ethical Investment Association www.ethicalinvestment.org.uk
UK Social Investment Forum www.uksif.org • 020 7405 0040
Ethical Investment Research Service www.eiris.org • 020 7840 5703

For more information about IFAs in general, contact IFAP, the UK association that promotes the profession. Their website lets you search for advisers in your area, and they can give some information about specialists in ethical matters:

IFAP www.ifap.org.uk • 0117 971 1177

Other financial services
Credit cards, insurance and mortgages

Credit cards

None of the major credit card providers are known for being particularly green or ethical companies, but there are a wide range of charity (or "affinity") credit cards available, which support environmental and other causes. Usually, when you sign up for such a card, the issuing company will make a lump payment (often in the region of £25) to the relevant charity. After that, they'll make a small donation each time you spend a certain amount on the card – for example, 25p or 50p donated for each £100 spent. Some charity cards, such as the American Express RED (pictured), which supports AIDS charities, give a full 1% of the money spent on the card to their associated cause.

Most charity cards have no fixed charges and offer perfectly competitive rates, so if you use a credit card, there's no reason not to sign up for one. Of course, shopping won't save the planet, and the items you purchase won't have a smaller carbon footprint just because you're using a credit card that supports an environmental group. But for goods you were going to buy anyway, there's no environmental disadvantage – and some environmental advantage – to using such a card.

Many of the high-street banks offer affiliate cards. For example, Barclaycard's Breathe card puts 50p for each £100 spent towards carbon-busting projects, and also offers a special low interest rate for purchases of public transport tickets. But you might prefer to use a green card from a more specifically ethical supplier – such as the Co-operative Bank, whose Think card supports deforestation charity Cool Earth and offers discounted rates on purchases from certain shops that the bank has designated as being environmentally friendly.

Insurance

Like pension companies, insurers have huge assets – it is estimated that at any one time the industry controls roughly 10% of the world's capital flows. Technically speaking this isn't consumers' money, but to all intents and purposes insurance funds are collectively owned by their policyholders – and, since the money is widely invested, that makes taking out insurance a bit like any other form of investing.

You might expect, therefore, that there would be a range of green and ethical insurance schemes on offer from companies operating screened investment policies. This is true for the life assurance side of the industry, but when it comes to the other side – the one that insures homes, contents, travel and the like – the options are surprisingly limited. Some companies claim to use their power as shareholders as a force for good, but investment screening is still basically nonexistent in the sector.

One reason for this is the interrelated structure of the industry. To be an actual insurer – rather than a small insurance firm selling policies underwritten by an insurer – requires an enormous amount of money. Even if you did have the required cash, you'd rely on a re-insurer to underwrite you, and there aren't any re-insurers with serious ethical investment policies. Also, according to some insiders in the industry, the culture of the whole sector is inherently at odds with the idea of anything beyond profits – even more so than in other financial industries.

Quite apart from what your insurance company is investing in, another issue is who and what insurance companies are willing to cover. Since many environmentally harmful projects wouldn't be feasible without insurance, the big companies collectively wield a major influence over what kinds of infrastructure projects do and don't go ahead. As such, they could be a force for good if they imposed industry-wide environmental or ethical standards.

For both these reasons, it's worth at least considering the two insurance companies with the best claims to being environmentally responsible.

CIS www.cis.co.uk • 08457 464 646

CIS have had a "Responsible Shareholder" approach for years and in 2005 became the first major insurance company to launch a fully fledged ethical policy. The principles in the policy largely match those of the Co-op Bank, though CIS only promises to promote those principles through engagement with companies rather than shunning certain sectors. On a broader level, CIS has flaunted its green credentials by cladding the support tower to its enormous office block in solar panels.

Naturesave Policies www.naturesave.co.uk • 01803 864 390

This small insurance intermediary covering home, contents, travel and business sells policies underwritten by Lloyd's of London. It claims to be deeply committed to sustainable development, and 10% of the premium made from each policy sold is put towards the Naturesave Trust, which funds environmental and conservation projects (this is taken from the profits rather than added to the policy price). The company also lobbies the insurance industry to spend money on dealing with environmental risks at their roots instead of paying out compensation once disasters happen.

Mortgages

Whereas it's quite clear what makes a bank or investment fund ethical (it attempts to exert an influence somewhere by selectively investing or by being an active shareholder), an "ethical mortgage" seems to mean different things to different people.

One consideration relates solely to interest-only mortgages. With these, instead of paying off part of the loan and part of the interest each month, you only pay back part of the interest, the rest of your payments typically being used to invest in a package, such as an endowment or ISA, that at the end of a fixed term will be used to pay off the mortgage. This is roughly the same as any other kind of investment, so the same ethical screening processes can be applied. Many companies now offer ethical endowments.

Another consideration is the ethical credentials of the lender. After all, during the lifespan of a mortgage you'll potentially be handing over tens of thousands of pounds' worth of profit to whoever lent you the money in the first place. So if, for example, you don't trust the green or ethical standards of the major banks, it would certainly be a logical step to avoid them for mortgages as well as for current accounts. On this basis, the typical green advice for mortgages is similar to that for banks: support the ethically enlightened Co-operative Bank or use a mutual building society, which will be unlikely to do anything more dodgy with the margin they make from the mortgage than pay it as interest to savers. Happily, mutuals very often offer a better deal anyway, so favouring them needn't mean bigger charges or monthly payments.

For more comprehensive information on mortgages, seek the help of a clued-up financial adviser (see p.302). One company that specializes in mortgages is the Ethical Investors Group.

Co-operative Bank www.co-operativebank.co.uk • 08457 212 212
As well as offering good rates on its CAT-standard mortgages, the Co-operative Bank, for each year that your mortgage with them exists, will offset a fifth of a typical home's greenhouse emissions. You might see this as a bit of a gimmick (it equates to around £8 per year in offset payments, which is nothing compared to the typical annual cost of a mortgage) but in the case of the Co-op you can rest assured that this offering is only part of a much broader commitment to green and ethical issues.

The Ecology Building Society www.ecology.co.uk • 0845 674 5566
This small mutual building society lends on the grounds of the building rather than the individual, giving mortgages to fund energy-efficient housing, renovation of derelict and dilapidated properties, small-scale and ecologically driven enterprise, and "low-impact lifestyles". Many lenders refuse to touch derelict properties, but the

Ecology, being small, can work one-to-one with borrowers to make sure their renovation project is financially sound. And if renovation is going well, and the building gaining value, the society may offer further funds for more improvements.

Ethical Investors Group www.ethicalmortgage.co.uk • 01242 539 848

Formed in 1989, this Cheltenham-based company (which donates half of its profits to charity) offers financial advice on ethical borrowing. It has grouped mortgage lenders into five categories according to their ethical positions and promises to help find the "best mortgage rate from the lender that you feel most comfortable with".

Norwich & Peterborough www.npbs.co.uk • 0845 300 6727

Norwich & Peterborough, another mutual building society, offers "Green" mortgages for new energy-efficient buildings or older dwellings in line for an eco-renovation. For each Green mortgage sold, the N&P will plant eight trees a year for five years, enough in theory to absorb the carbon dioxide produced by the property.

Carbon offsetting

Offsets and alternatives

Carbon offset schemes allow individuals and companies to invest in environmental projects around the world in order to balance out their own carbon footprints. The projects are usually based in developing countries and most commonly are designed to reduce future emissions. This might involve rolling out clean energy sources, distributing energy-saving devices like eco light bulbs, or purchasing and ripping up carbon credits (see p.34). Other schemes work by soaking up CO_2 directly from the air through the planting of trees.

Some people and organizations offset their entire carbon footprint while others aim to neutralize the impact of a specific activity, such as taking a flight. To do this, the holidaymaker or business person would visit an offset website, use the online tools to calculate the emissions of their trip, and then pay the offset company to reduce emissions elsewhere in the world by the same amount – thus making the flight "carbon neutral".

Offset schemes vary widely in terms of the cost, though a fairly typical fee

A typical carbon offset website, with tools to calculate the emissions and offset cost of various activities would be around £8/$12 for each tonne of offset CO_2. At this price, a typical British citizen would pay £45 to neutralize a year's worth of gas and electricity, while a return flight from London to San Francisco would clock in at around £20 per ticket.

Increasingly, many products are also available with carbon neutrality included as part of the price. These range from books about environmental topics (such as *The Rough Guide to Climate Change*, whose paper, print and distribution were offset by the publisher) through to high-emission cars (Land Rover include offsets for a certain amount of mileage in the price of each new energy-inefficient car).

Over the past few years, carbon offsetting has become increasingly popular, but it's also become – for a mixture of legitimate and less legitimate reasons – increasingly controversial. This chapter takes a quick look at both sides of the argument.

Is the whole concept of offsetting a scam?

Ironically, perhaps, the most common criticisms of offsetting relate to the planting of trees. These concerns are perfectly valid (see box on p.310), but in truth most of the best-known carbon offset schemes have long-since switched from tree planting to clean-energy projects. For example, Climate Care distribute efficient cooking stoves to families in Mexico and Honduras. Energy-based projects such as these are designed to make quicker and more permanent savings than planting trees, and, as a bonus, to offer social benefits. Efficient cooking stoves, for instance, can help poor families save money on fuel and improve their household air quality – a very real benefit in many developing countries.

Even in the case of energy-based schemes, however, many people argue that offsetting is unhelpful – or even counterproductive – in the fight against climate change. One such person is environment journalist George Monbiot, who famously compared carbon offsets with the ancient Catholic Church's practice of selling indulgences: absolution from sins and reduced time in Purgatory in return for financial donations to the Church. Just as indulgences allowed the rich to feel better about sinful behaviour without actually changing their ways, carbon offsets allow us to "buy complacency, political apathy and self-satisfaction", Monbiot claims. "Our guilty consciences appeased, we continue to fill up our SUVs and fly round the world without the least concern about our impact on the planet … it's like pushing the food around on your plate to create the impression that you have eaten it."

cheatneutral.

Helping you because you can't help yourself

about   our projects   become a project   offset your cheating   press/contact   film   small print

Our projects

We at Cheatneutral are committed to saving relationships, promoting fidelity, and making people feel good about themselves. Our projects have a great success rate.

Steve and Lisa

Steve and Lisa met while on holiday in Spain, and quickly fell head over heels for each other. That Christmas, at his office party, Steve got drunk and unavoidably repeatedly cheated on Lisa with Cheri, a co-worker. He paid Cheatneutral just £2.50 and we invested his money in Alex, a single man with no prospect of finding a partner. In return for the payments, Alex agreed to remain single.

A similar if more humorous point is made by the spoof website CheatNeutral.com, which parodies carbon neutrality by offering a similar service for infidelity. "When you cheat on your partner you add to the heartbreak, pain and jealousy in the atmosphere", the website explains. "CheatNeutral offsets your cheating by funding someone else to be faithful and not cheat. This neutralizes the pain and unhappy emotion and leaves you with a clear conscience."

CheatNeutral may be tongue-in-cheek but the indulgence and cheating analogies have both become defacto arguments against carbon offsetting. But do the comparisons stand up? Not according to David Roberts, staff writer at Grist. "If there really were such a thing as sin, and there was a finite amount of it in the world, and it was the aggregate amount of sin that mattered rather than any individual's contribution, and indulgences really did reduce aggregate sin, then indulgences would have been a perfectly sensible idea", Roberts argues. "The comparison is a weak and transparent smear, which makes me wonder why critics rely so heavily on it."

And what about the claim that people use offsetting as a way to avoid changing their eco-unfriendly ways? This is nonsense, too, according to the offset schemes, which claim that most of their customers are also taking steps to reduce their emissions directly. A report from Britain's National Consumer Council and Sustainable Development Commission agreed with this perspective: "a positive approach to offsetting could have public resonance well beyond the CO_2 offset, and would help to build awareness of the need for other measures."

Planting trees: does it help?

Ronald Reagan said a good many baffling things during his eight-year presidency, but none quite matches his 1981 claim that "trees cause more pollution than automobiles". It doesn't take an environmental science degree to detect that there might be something wrong with that statement. That said, over the years, and especially since the emergence of carbon offsetting, the environmental benefits of trees – and in particular the planting of new ones – have been challenged and debated. So what's the truth? Will planting a tree in your garden, or paying companies to plant saplings on your behalf, help tackle climate change?

The fact that trees breathe in CO_2 is not in question. They need the carbon to grow (they're largely made of the stuff), and you can observe the effect playing out in the world's atmosphere each year. As the graph shows, even as the level of CO_2 in the air continues its upward climb due to fossil fuel burning, the precise concentration actually falls a bit each year during the growing season of the northern hemisphere, where most of the world's trees and vegetation exists.

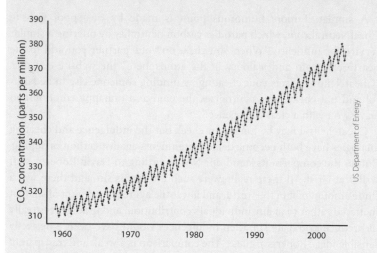

That basic science aside, there are two frequently cited downsides to using trees to soak up carbon. First, trees can take a relatively long time to grow, so it's not a very rapid solution. Second, and more fundamentally, trees eventually die, at which point they rot – or get burned – allowing much of their stored carbon back into the air. For this reason, contrary to popular belief, a stable forest doesn't actually absorb a huge amount of carbon each year: for each tree growing and sucking in CO_2 another is rotting and returning much of its carbon to the atmosphere.

Of course, planting trees in areas where there weren't any previously *will* soak up CO_2, though the benefit won't be permanent unless each tree that dies is replaced by another. This would be hard enough to guarantee over decades or centuries even if the climate were stable. But if, as most experts expect, global warming increases temperatures by a few degrees over the coming century, then climate change could, somewhat ironically, kill those very trees planted to offset the emissions that helped cause the climate change in the first place. In this case, the carbon would be released back into the atmosphere, removing any climate benefit that the trees offered while living.

Furthermore, although trees absorb CO_2 as they grow, that doesn't necessarily mean they always reduce global warming overall. Scientists now think, for example, that in northern temperate and arctic areas, some trees have an overall warming effect because their dark colour tends to absorb more sunlight and reflect less back into space than lighter surfaces such as snow would do. (It's the same phenomenon as the extra heat you feel when wearing dark clothes in summer.) By contrast, tropical rainforests cool the air by trapping water and letting it slowly evaporate.

Planting trees in the tropics, then, is the most effective option. This is doubly true because tropical trees tend to grow fast and therefore absorb carbon fairly quickly. On the other hand, most tropical regions are in developing countries, and some critics have described schemes by Westerners to plant trees there as a type of "green colonialism". It's wrong, such commentators argue, for Western organizations to determine how land is used overseas.

Another, totally separate, tree-planting controversy sprung up in early 2006 when a team of scientists led by Frank Keppler of the Max Planck Institute made the surprise discovery that trees and other plants emit small quantities of methane, a greenhouse gas that's shorter lived but much more powerful than CO_2. The scientists estimated that plants might account for 10–30% of current methane emissions. If true, that's a huge overall impact but the methane emitted from a single tree doesn't begin to outweigh the CO_2 it soaks up – it just "reduces the overall benefit … by a fraction", according to Yadvinder Malhi of Oxford University.

All in all, despite the controversies, planting a tree where there wasn't one before *is* likely to help fight climate change – at least in the short and medium terms. So schemes to add trees in cities, parks, gardens and elsewhere should be welcomed. But trying to offset the emissions of a flight or anything else by planting trees is not necessarily legitimate, especially if they're being planted in a cold country.

Perhaps a more important question is how we protect the trees that are already standing – and in particular the world's tropical rainforests. As mentioned in chapter one, the destruction of rainforests accounts for nearly a fifth of recent man-made greenhouse emissions: more than the US or China or the EU. One way to help reduce deforestation is to join a charity such as Cool Earth, described on p.315.

Ultimately, the question of whether the concept of offsetting is valid must come down to the individual. If you offset to assuage guilt and to make yourself feel better about high carbon activities such as flying, that can't be good. If you offset as part of cutting your footprint, or to incentivize yourself to be greener (after all, the less you emit, the less it will cost you to go carbon neutral) then that can't be bad – especially if the offset projects offer extra benefits such as poverty reduction in the developing world.

Do offset projects actually deliver the carbon benefits they promise?

Arguments about guilty consciences aside, the key issue for anyone who does want to offset is whether the scheme you're funding actually achieves the carbon savings promised. This boils down not just to the effectiveness of the project at zapping CO_2 or avoiding future emissions. Effectiveness is important but not enough. You also need to be sure that the carbon savings are additional to any savings which might have happened anyway.

Take the example of an offset project that distributes low-energy light bulbs in a developing country, thereby reducing energy consumption over the coming years. The carbon savings would only be classified as additional if the project managers could demonstrate that, for the period in which the carbon savings of the new light bulbs were being counted, the recipients *wouldn't* have acquired low-energy bulbs by some other means.

The problem is that it's almost impossible to prove additionality with absolute certainly, as no one can be sure what will happen in the future, or what would have happened if the project had never existed. For instance, in the case of the light-bulb project, the local government might start distributing low-energy bulbs to help reduce pressure on the electricity grid. If that happened, the bulbs distributed by the offset company would cease to be additional, since the energy savings would have happened even if the offset project had never happened.

Partly because of the difficulty of ensuring additionality, many offset providers guarantee their emissions savings. This way, if the emissions savings don't come through or they turn out to be "non-additional" (in the case of the government giving out low-energy light bulbs), then the provider promises to make up the loss via another project.

As the offset market grows, some offset companies have enough capital to invest in projects speculatively: they fund an offset project and then sell the carbon savings once the cuts have actually been made. This avoids the difficulty of predicting the future – and also avoids the claim that a carbon cut made some years in the future is worth less than a cut made now.

These kinds of guarantees and policies provide some reassurances, but do they mean anything in the real world? Without actually visiting the offset projects ourselves, how can individuals be sure that the projects are functioning as they should?

To try and answer these questions, the voluntary offset market has developed various standards, which are a bit like the certification systems used for fairly traded or organic food. Industry insiders generally say that the best of these schemes are the Voluntary Gold Standard (VGS) and the Voluntary Carbon Standard (VCS). VGS-certified offsets are audited according to the rules laid out in the Kyoto Protocol and must also show social benefits for local communities. The VCS, meanwhile, aims to be just as rigorous but without being too expensive or bureaucratic to set up, thereby allowing a greater range of innovative small-scale projects.

Offsets with these standards offer extra credibility, but that still doesn't make them watertight. Heather Rogers, author of *Made in the Shade*, visited a number of offset schemes in India and found all kinds of irregularities with the projects there. One VGS-certified biomass power plant refused to allow her around, though staff there reported a number of concerns such as trees being chopped down and sold to the plant, which was designed to run on agricultural wastes.

Even if offset projects *do* work as advertised, some environmentalists argue that they're still a bad idea. If we're to tackle climate change, they

The price of offsetting

Many people are confused by the low prices of carbon offsets. If it's so bad for the environment to fly, can a few pounds really be enough to counteract the impact? The answer is that, at present, there are all kinds of ways to reduce emissions very inexpensively. After all, a single low-energy light bulb, available for just £1, can save 250kg of CO_2 emissions over its lifetime – equivalent to flying return from London to Berlin. That's not to say that offsetting is valid, or that plugging in a low-energy light bulb makes up for flying to Berlin. The point is simply that the world *is* full of inexpensive ways to reduce emissions. In theory, if enough people started offsetting, or if governments started acting seriously to tackle global warming, then the price of offsets would gradually rise, as the low-hanging fruit of emissions savings – the easiest and cheapest "quick wins" – would get used up.

argue, the projects being rolled out by offset companies should be happening anyway, funded by governments around the world, while companies and individuals reduce their carbon footprints directly. Only in this way – by doing everything possible to make reductions everywhere, rather than polluting in one place and "offsetting" in another – does the world have a good chance of avoiding runaway climate change.

Offsetting companies

If you do choose to offset, it also makes sense to at least opt for providers which are respected in the field. To help confused consumers, an organization called Clean Air Cool Planet (cleanair-coolplanet.org) assessed thirty providers against a wide range of criteria in December 2006 and named the following eight companies as "top performers".

AgCert drivinggreen.com
Atmosfair atmosfair.de
CarbonNeutral Company carbonneutral.com
Climate Care jpmorganclimatecare.com
Climate Trust climatetrust.org
CO2Balance co2balance.com
NativeEnergy nativeenergy.com
MyClimate my-climate.com

Offsetting for countries

The concept of offsetting and additionality have their roots in the Clean Development Mechanism (CDM), the carbon-trading system built into the Kyoto Protocol. The CDM allows developed countries to pay for carbon cuts in developing countries instead of making more expensive emissions reductions at home. A rich country struggling with its Kyoto targets might, for example, fund a hydroelectric station in China or India.

As with small-scale offsetting, the CDM has been a source of much controversy over the years. Though some projects have worked well, the additionality of others has been questioned, and some have been shown to be almost laughably expensive for the carbon savings delivered.

It remains to be seen how big a role nation-to-nation offsetting will be given in the global climate deal due to follow on from Kyoto in 2012. Green groups tend to argue that rich countries should make all their emissions reductions at home – not by paying other countries.

Alternatives to offsets

The debate about the validity of offsetting is bound to run and run. In the meantime, some companies and individuals are switching from traditional offsets to alternative charity-like schemes that promise big environmental benefits rather than carbon neutrality. These include the following.

Sandbag www.sandbag.co.uk

This small London-based charity aims to interact with the EU's emissions trading system which allocates tradable CO_2 permits to carbon-intensive industries. Sandbag uses supporters' money to buy these permits and tear them up, hence reducing the amount of CO_2 that companies are legally allowed to produce. Sandbag also acts as a campaign group, lobbying companies that have been allocated more CO_2 permits than they need to surrender them rather than sell them, thereby reducing the amount of CO_2 entering the atmosphere.

Cool Earth www.coolearth.org

While people debate the exact carbon benefits of planting new trees, everyone agrees that it's critical that the world stops the destruction of existing forests. Cool Earth takes donations and uses them to protect critically endangered rainforest. In some cases, the charity buys the land and donates it to a local trust, with local people employed as forest stewards; in other cases, when forest-owning tribes have been offered money for logging rights in their land, the charity offers a similar amount in return for forest protection. Cool Earth doesn't see itself as an offset scheme but points out that protecting a single acre – at a cost of around £75 – can avoid more than two hundred tonnes of CO_2 emissions. The same money spent via a traditional offset scheme would save something closer to ten tonnes.

The Converging World www.theconvergingworld.org

Rather than offering offsets, this charity accepts "money being given in acknowledgement of a carbon impact" and uses it to fund renewable energy projects in India. The cash generated from these schemes – typically "more than twice what is put in" – is used to fund development projects.

Index

Index

T

U